Migration and Worker Fatalities Abroad

DOI: 10.1057/9781137451187.0001

Other Palgrave Pivot titles

Debra Reddin van Tuyll, Nancy McKenzie Dupont and Joseph R. Hayden: **Journalism in the Fallen Confederacy**

Michael Gardiner: **Time, Action and the Scottish Independence Referendum**

Tom Bristow: **The Anthropocene Lyric: An Affective Geography of Poetry, Person, Place**

Shepard Masocha: **Asylum Seekers, Social Work and Racism**

Michael Huxley: **The Dancer's World, 1920–1945: Modern Dancers and Their Practices Reconsidered**

Michael Longo and Philomena Murray: **Europe's Legitimacy Crisis: From Causes to Solutions**

Mark Lauchs, Andy Bain and Peter Bell: **Outlaw Motorcycle Gangs: A Theoretical Perspective**

Majid Yar: **Crime and the Imaginary of Disaster: Post-Apocalyptic Fictions and the Crisis of Social Order**

Sharon Hayes and Samantha Jeffries: **Romantic Terrorism: An Auto-Ethnography of Domestic Violence, Victimization and Survival**

Gideon Maas and Paul Jones: **Systemic Entrepreneurship: Contemporary Issues and Case Studies**

Surja Datta and Neil Oschlag-Michael: **Understanding and Managing IT Outsourcing: A Partnership Approach**

Keiichi Kubota and Hitoshi Takehara: **Reform and Price Discovery at the Tokyo Stock Exchange: From 1990 to 2012**

Emanuele Rossi and Rok Stepic: **Infrastructure Project Finance and Project Bonds in Europe**

Annalisa Furia: **The Foreign Aid Regime: Gift-Giving, States and Global Dis/Order**

C. J. T. Talar and Lawrence F. Barmann (editors): **Roman Catholic Modernists Confront the Great War**

Bernard Kelly: **Military Internees, Prisoners of War and the Irish State during the Second World War**

James Raven: **Lost Mansions: Essays on the Destruction of the Country House**

Luigino Bruni: **A Lexicon of Social Well-Being**

Michael Byron: **Submission and Subjection in Leviathan: Good Subjects in the Hobbesian Commonwealth**

Andrew Szanajda: **The Allies and the German Problem, 1941–1949: From Cooperation to Alternative Settlement**

DOI: 10.1057/9781137451187.0001

palgrave▶pivot

Migration and Worker Fatalities Abroad

AKM Ahsan Ullah
Associate Professor, University Brunei Darussalam, Brunei

Mallik Akram Hossain
Professor, Jagannath University, Bangladesh

and

Kazi Maruful Islam
Associate Professor, University of Dhaka, Bangladesh

palgrave
macmillan

DOI: 10.1057/9781137451187.0001

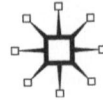

First published 2015 by
PALGRAVE MACMILLAN

Palgrave Macmillan in the UK is an imprint of Macmillan Publishers Limited,
registered in England, company number 785998, of Houndmills, Basingstoke,
Hampshire RG21 6XS.

Palgrave Macmillan in the US is a division of St Martin's Press LLC,
175 Fifth Avenue, New York, NY 10010.

Palgrave Macmillan is the global academic imprint of the above companies
and has companies and representatives throughout the world.

Palgrave® and Macmillan® are registered trademarks in the United States,
the United Kingdom, Europe and other countries.

ISBN: 978–1–137–45119–4 EPUB
ISBN: 978–1–137–45118–7 PDF
ISBN: 978–1–137–45117–0 Hardback

A catalogue record for this book is available from the British Library.

A catalog record for this book is available from the Library of Congress.

www.palgrave.com/pivot

DOI: 10.1057/9781137451187

Contents

List of Illustrations

Figures

Map

Tables

DOI: 10.1057/9781137451187.0002

DOI: 10.1057/9781137451187.0002

Acknowledgements

We would like to extend our thanks to the participants of the study, without whom this study and manuscript could not have been completed. Most of the participants of this study are the family members of workers who died while employed abroad. We are keenly aware of their loss and sacrifice and are grateful for their participation and help in preparing this painful study.

I would like to share our greatest appreciation to Jenny Fan of the University of California, Irvine (UCI) for her persistent cooperation and support throughout this project period and beyond. She has always been extremely prompt in responding to our questions and concerns. We owe thanks to Professor Bill Maurer, Dean of Social Sciences, the University of California, Irvine, for his advice and support at various levels.

We are indebted to the Institute for Money, Technology and Financial Inclusion (IMTFI), University of California, Irvine, for granting the funds for this research. This allowed us a great opportunity to shed light on one of the most crucial topics of study in the arena of labour migration.

Our thanks and appreciations also go to our colleague, Md Mizanur Rahman, National University of Singapore (NUS) for his continued cooperation and valuable advice. We must also thank the research assistants involved in the research project. They have been extremely enthusiastic during the process of collecting data. Their precious assistance is greatly acknowledged. We wish to acknowledge also the help provided by Md Amirul Islam, Cairo University,

DOI: 10.1057/9781137451187.0003

for his support in reference checking. We thank Joanna Safford for her extraordinary editorial support.

It would not have been possible to give the research the shape of a book without the kind support and help of many individuals and organizations. We would like to express our gratitude to the people who have been instrumental in the successful completion of this project. Last but not least, we would like to thank all our research assistants for aiding us in the process of participant selection and interviews.

DOI: 10.1057/9781137451187.0003

List of Abbreviations

AHRC	Asian Human Rights Commission
AIDS	Acquired Immunodeficiency Syndrome
APC	Asia-Pacific Consultations on Refugees, Displaced Persons and Migrants
ASEAS	Association of Southeast Asian States
BAIRA	Bangladesh Association of International Recruiting Agencies
BBS	Bulletin Board System
BMET	Bureau of Manpower Employment and Training
BOESL	Bangladesh Overseas Employment and Services Ltd.
CARAM	Coordination of Action Research on AIDS and Mobility
CEDAW	Convention on the Elimination of All Forms of Discrimination against Women
DOLE	Department of Labour and Employment
FIDH	International Federation for Human Rights
G2G	Government to Government
GCC	Gulf Cooperation Council
HDI	Human Development Index
HDR	Human Development Report
HIV	Human Immunodeficiency Virus
HKSAR	Hong Kong Special Administrative Region
HRW	Human Rights Watch
IFHR	International Federation of Human Rights
ILM	International Labor Migration
ILO	International Labor Organization
IMA	International Migrants Alliance

DOI: 10.1057/9781137451187.0004

IMTFI	Institute For Money, Technology and Financial Inclusion
IOM	International Organization for Migration
MEWOE	Ministry of Expatriates' Welfare and Overseas Employment
KSA	Kingdom of Saudi Arabia
MOL	Ministry of Labour
MOSAL	Ministry of Social Affairs and Labour
MOU	Memorandum of Understanding
MWPS	Migrant Workers Protection Society
NGOs	Non-Governmental Organizations
OECD	Organisation for Economic Cooperation and Development
OHCHR	Office of the High Commissioner for Human Rights
OIZs	Organized Industrial Zones
OWWA	Overseas Workers Welfare Administration
PDOS	Pre-Departure Orientation Seminar
POEA	Philippines Overseas Employment Administration
RCP	Regional Consultative Process
SLBFE	Sri Lanka Bureau of Foreign Employment
SSWA	Sanitary Supply Wholesaling Association
TTC	Technical Training Centers
UN	United Nations
UNB	Union National Bank
UNDP	United Nations Development Programme
UNESCO	United Nations Educational, Scientific and Cultural Organization
UNFPA	United Nations Fund for Population Activities
UNHCR	United Nations High Commissioner for Refugees
UNICEF	United Nations International Children's Emergency Fund
USDS	United States Department of State
WHO	World Health Organization
WTO	World Trade Organization

DOI: 10.1057/9781137451187.0004

palgrave▸**pivot**

www.palgrave.com/pivot

1

Introduction: Understanding Migration and Fatalities

Abstract: *Population migration across the world has not only brought blessings but also pain to many people. Drawing on the recent migration theories and debates the chapter lays out the theoretical framework to understand the relation between migration and fatalities. The chapter explains how a set of variables including workers' access to basic services, level of stress, nature of politics and policy contribute to increased fatalities of the migrants. With the data from South Asia, the authors explore causes of health hazards and fatalities of the migrant. The Institutionalised Dependency Trap (IDT) is also elaborated upon in this chapter in order to explain why and how migrant casualties happen, especially in the Middle East and Gulf countries. Attention has also been drawn to the policies of the receiving countries that serve only to intensify the dependency trap for the mostly South Asian migrants. The chapter presents the methodology and concludes with the organization of chapters in the book.*

Ullah, AKM Ahsan, Mallik Akram Hossain and Kazi Maruful Islam. *Migration and Worker Fatalities Abroad.* Basingstoke: Palgrave Macmillan, 2015. DOI: 10.1057/9781137451187.0005.

> Every 15 seconds, a worker dies from a work-related accident or disease. Every 15 seconds, 160 workers have a work-related accident – International Labour Organization (ILO).

Population migration has not brought blessings equally to everyone. For some countries, migration has brought economic vibrancy while for others skill-drain has caused severe economic strain. At an individual level, some have gained immensely while others have lost miserably. What is this loss about? It comes in myriad shapes, forms, scales and patterns. The proliferation of people on the move globally, and the accompanying financial transfers we call remittances, have brought migration issues to the fore in the global development agenda.

Population mobility is as old as human history. Nonetheless, very little is known about the dynamics of its prehistory. Only the trajectory of gypsy mobility has been well documented, though clear reasons for their nomadic tradition remain largely unknown (Ullah, 2010a, 2011, 2011a; Kutasi, 2005). Globalization has triggered a process of worldwide integration, which was abetted by the massive influx not only of ideas, goods and capitals across borders, but also of people. The transfer of people has been less intense than any of the other elements, since it faces greater political restrictions and is subject to more explicit and implicit barriers (Alonso, 2011). The general assumption is that the introduction of modern communication technology, better economic opportunities, and secured life opportunities have enticed people to migrate. Therefore, peoples' awareness about opportunities and networking possibilities has increased over the last 40 years as a consequence of the ongoing process of globalization. However, global, structural socio-economic characteristics such as income disparities, demographic imbalances and labour market discrepancies also rose up (IOM, 2008).

Different schools of thought offer different explanations as to why some people emigrate and why some do not. According to economic schools of thought, migration is a function of demand and supply. Demographers argue that it is a result of population pressure, that is, countries with surplus labour send and countries with deficient labour receive. However, the prior theory applies to the case of Bangladeshi emigration (Ullah, 2009). Within this "labor surplus-resource deficit" syndrome, the rate of labour migration is increasing rapidly. For a country like Bangladesh, with a 33.8 per cent underemployment rate, better

DOI: 10.1057/9781137451187.0005

job opportunities abroad could explain the increasing outward flow of migrant workers (BMET, 2010).

In the past four decades there has been a dramatic upward trend in both skilled and unskilled labour migration. In 1965 there were only 75 million international migrants; the figure of migration has shot up to about one billion at present (both national and international) (Ullah, 2010a, 2011; IOM, 2011). The total volume of international migrants represents almost 3 per cent of the global population (Susan, 2001; IOM, 2010). The mere numbers do not present the whole of the issue; the social and political dimensions of migration make it a matter of vital importance, "migration involves people, agents bearing life plans, dreams and frustrations, hopes, interests and cultures" (Alonso, 2011). Though international migration is a central feature of the present international economy, it has never received the same amount of attention lavished on the theory of international capital movements. The mass movement of migrants with few skills has generated a large descriptive literature. In more recent years, however, the problems of international migration have been analysed more and more on a theoretical and empirical level. No organization at the global level is currently responsible for systematically monitoring the number of deaths that occur.

In recent years, policymakers, researchers and academics have given growing attention to migration issues, which has helped discourse flourish on both the negative and positive effects of migration to a country's development. It is a well-researched phenomenon; however, most of the works have focused on one specific aspect, that is, the migration (economic) development nexus, ignoring the very fact that migration is a multidimensional development issue. More precisely, no significant amount of research is found on the fate of the migrant workers in terms of their social and cultural well-being, their survival and coping strategies, and most importantly, the status of rights and entitlements they are allowed to enjoy in the destination country. Clearly, there is a dearth of research on the numbers of migrant workers who die abroad, the cause of their deaths and the manner of how their families are compensated. This research investigates the reasons for the deaths of migrant workers abroad. It also studies the impact that a migrant's demise can bring to the left-behind family.

The incidences of death are escalating in numbers dramatically, and are rightly a growing cause of concern. A recent study reports that the average number of dead bodies returned from abroad to Bangladesh per day is about eight (*Daily Naya Diganta*, 18 July 2009). Between 2003 and 2012, a staggering 15,752 dead bodies arrived from abroad, mostly from

DOI: 10.1057/9781137451187.0005

the Middle East and Malaysia. This is without doubt a horrific picture (Shariful, 2012; Ullah and Hossain, 2013).

The primary causes of the deaths, as documented by death certificates, are cardiac arrests, workplace accidents, road accidents, mental stress and other illnesses. Cardiac arrest recorded the highest number of casualties. Death certificates claim that almost all of the Bangladeshi workers who returned home in coffins between 1 January and 9 May in 2009, died of cardiac arrest in Middle Eastern and Southeast Asian countries (Charles, 2009). Relevant ministers as well were heard to say that these were natural deaths. However, most family members who were left behind did not accept the stated reasons for their death. Of course, the rate is more than 20 times higher than the national rate. Some experts now believe that for every dead body discovered, there are at least two others that are never recovered.

Media outlets continuously report that Bangladeshis live and work in precarious conditions in the Middle East and the Southeast Asian countries. We are reminded that in Saudi Arabia, six Bangladeshis were burned alive in their workplace (Prothom Alo, 2009). They were not able to escape. There was no "exit" point. They had surrendered themselves to death helplessly. This incident is an example of how significant abuses are inflicted on the migrant workers. It illustrates how abject is the work environment. The notorious "3-D jobs" – dangerous, dirty and demeaning – elucidate the overall standard of labour safety (Palma, 2008). Extremely low wages, unpaid wages for months, and the confiscation of passports as "security" to prevent workers from quitting are part of the realities of the Bangladeshi migrants. Despite the miseries subjected on their nationals, Bangladesh's efforts to end the abuses have been insignificant.

The motivation for migration has a direct association with the wage differential, but, conversely, has an inverse ratio to the distance (Kutasi, 2005). Quite surprisingly, psychological and opportunity costs are in general not taken into account when they make their decision to migrate.

A framework of analysis

In the current globalized world, patterns and dynamics of migration have become highly varied, complex and context specific (Collinson, 2009). Explaining and understanding this complex process, thus, calls

DOI: 10.1057/9781137451187.0005

for multidisciplinary approaches. While classical and contemporary theories of migration explain the factors that shape migration behaviour, a detailed framework has yet to be developed to produce plausible explanations of what happens to the migrant workers, and why and how it happens after migration. Notwithstanding the fact that by the end of the 20th century all industrially developed countries, all of the oil-rich Middle Eastern countries and a certain number of newly industrialized countries, have become countries of immigration, all of them have not been able to put well-defined policies, legal frameworks and institutions in place to deal with the incoming migrants. To put simply, not all of the destination countries are equally institutionally equipped to handle the situation of the migrants, especially the migrant workers. While some countries, for example, United States and Canada, have well-developed legislative and policy framework, countries like Kuwait and Saudi Arabia have deliberately avoided establishing any such institutional set-up (Massey, 2009).

The fate of migrant workers is continuously being shaped and governed by the social, political and economic forces, the legislative framework and institutional configurations in the receiving countries. The degree and pattern of interactions among these forces and institutions is largely determined by the state. And the state does so through its immigration policy. Massey argues that the immigration policy is the "outcome of a political process through which competing interests interact within bureaucratic, legislative, judicial and public arenas to influence the flow of immigrants" (Massey, 2009, p. 31). Drawing upon the point Massey made, we argue immigration policy not only controls or regulates the "flow of immigrants", but also influences the fate of the immigrants in the receiving country. After reviewing Massey (2007, 2009), Collinson (2009), Massey et al. (1993) and King (2012) and other evidence, we propose the following analytical framework to explain the condition of migrant workers in the receiving countries:

1 The condition of migrant workers in general and in particular the degree of fatalities of migrant workers can be viewed as a political–economic outcome in the receiving country;

2 The immigration policy of the receiving country directly influences the flow and fate of migrant workers;

3 The immigration policy reflects competing interests, attitudes and perception of major social, political and economic groups in the

DOI: 10.1057/9781137451187.0005

country, towards the immigrants. The fatalities of migrant workers, therefore, could be viewed as an outcome of social, political and economic stresses;

4   The level of social stress depends on two factors: firstly, the pattern of social relationship a migrant worker is able and allowed to build in the receiving society and secondly, the degree of access to social services such as housing, health care and psycho-social counselling. The level of political stress depends on the range of political and legal rights the migrant workers are granted to enjoy, for example, the right to form a trade union, the right to protest. Similarly, the level of economic stress depends on a couple of factors related to their livelihood, for example, the amount and regularity of the earnings, the right and opportunity to change their job and the status of their working conditions. In summary, the level of stress and number of fatalities are positively correlated; the higher the stress level, the higher the number of fatalities.

With reference to the condition of migrant workers in the receiving country, this framework allows us to refocus our attention on the politics of migration. In the migration politics the state plays the central role. The main function of the state is to formulate the migration policy, law and regulations to manage the transnational inflow of migrant workers and also the conditions they work and live in. In terms of the nature of the influence on migrant workers' work and living conditions in receiving countries the immigration policy could be categorized in two policies: repressive and progressive. Under the repressive immigration policy regime the migrant workers are allowed to enjoy limited access to social, political and economic opportunities and services, thus the level of stress goes high and eventually contributes to increased fatalities. Under the progressive policy regime migrant workers are allowed to live and work in more humane conditions with their rights ensured. Because of this humane and relatively liberal social environment their levels of stress reduce and the likelihood of fatalities gets decreases.

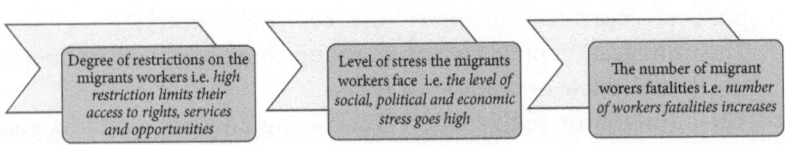

FIGURE 1.1   *The relation between workers' access, stress and fatalities*

DOI: 10.1057/9781137451187.0005

The nature of the immigration policy and worker fatalities is arguably associated. However, the numbers of migrant worker fatalities substantially vary from country to country and this is because the policies vary. And the policies vary because there is a marked contrast between the countries in terms of their system of governance and politics. In other words, the nature of the policy is directly associated with the system of governance and politics in the receiving country.

Building on Massey's (2009) propositions, we argue the nature of the governance system determines the nature of the policy. Broadly speaking, a system of governance can be categorized as democratic and non-democratic. These categories are developed based on a maximalist conception of democratic governance. The term "democratic governance" is perceived as a system of politics and government that promotes rule of law, independent judiciary, free, fair and competitive election, voice and accountability and observance of human rights. The dynamic interrelation among these five factors creates a continuum of governance. As argument flows, the countries that score high in the continuum take relatively progressive immigration policy. Meaning, in these countries, the migrants can enjoy civic, social and political rights. For example, they are granted legal access to housing, health care, family reunification and social benefits. For any kind of violation of rights, the migrant workers are entitled to go to court where their rights are constitutionally protected. In these countries the migrant workers are also allowed to form their own social organization. In this situation the workers are allowed to enjoy working life with less stress; thus, the risks of accidents and health hazards appear to be relatively low. The numbers of fatalities among them, therefore, do not increase alarmingly. Canada, the United Kingdom, Australia and the Scandinavian countries fall under these

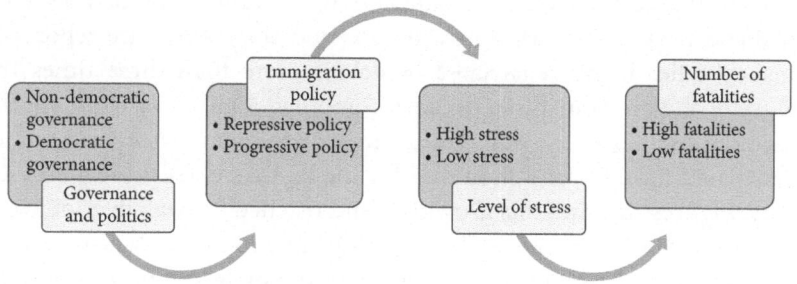

FIGURE 1.2   *Linkage between politics, policy and fatalities*

DOI: 10.1057/9781137451187.0005

categories. On the other extreme, there are countries featured with no or limited constitutional protection of social and political rights, authoritarian non-representative government, a poor rule of law and human rights record. In these countries the migrant workers are denied their access to social services like health care and their right to appeal to decisions about their status. Saudi Arabia, Kuwait and countries in the Gulf are some examples of this category. As Massey mentioned, none of the Gulf States recognizes the right to asylum or allows residence without a job. The migrant workers in these countries are explicitly excluded from their social and political structures (Massey, 2009). As a result of denial of the rights, lack of access to services and exclusion from political and social entitlements happen to the migrant workers. For them, hence, the level of stress goes high and eventually contributes to an increased number of fatalities.

In the following chapters of this book the above arguments will be analysed in the case of South Asian migrant workers with a special focus on the Bangladeshi workers working abroad.

Migration from Bangladesh and fatalities

There are no estimates of how many migrant workers die, how many are injured and how many become crippled while abroad every year. There is no body of research, or well-sourced estimate of these crucial figures. The extra legal status of many workers, the high level of abuse, and secrecy make it a virtually impossible task to estimate the death counts, and to segregate the death cases by cause. Now is the time to have a global, regional and national study determining the cause of deaths of migrant populations (outside their country of origin).

According to BMET, 23,641 Bangladeshi migrants have died between 1976 and 2013 (Table 1.1). Between 2003 and 2009, there are reports of 10,569 bodies being repatriated, which is more than three times the figure of the previous three decades combined. One can clearly see the alarming trend over the past decade; in 2005 the number of dead bodies repatriated ticked over a thousand for the first time, reaching a horrific 1,248. Figures showing this growing trend reached its peak in 2008, with 2,237 dead bodies received. The figures from 2009 to 2013 also tend to increase alarmingly and in 2013 represent the highest numbers of dead bodies recorded for the first time in the last decade. Table 1.1 shows that

DOI: 10.1057/9781137451187.0005

TABLE 1.1   *Number of dead bodies arrived in Bangladesh*

Month/Year	Number of death
1976–2002 (aggregated)	3,613
2003	906
2004	788
2005	1,248
2006	1,402
2007	1,673
2008	2,237
2009	2,315
2010	2,299
2011	2,235
2012	2,383
*2013	2,542
Total	23,641

Note: *The Prothom Alo, 10 February 2014.
Source: Compiled from BMET (2010).

the number of dead migrants surprisingly jumped to over 2000 from the year 2008, which continues to rise until 2013. It is believed that many more deaths remain unreported.

Migration pattern and trend

Very recently (end of September 2014) the International Organization for Migration (IOM) released data on migrant fatalities across land and sea. With a count surpassing 40,000 victims since 2000, the IOM calls upon all the world's governments to address this as the epidemic of crime and victimization. The DG of the IOM says "It is time to do more than count the number of victims. It is time to engage the world to stop this violence against desperate migrants." However, this is a publication from the IOM based on fatalities of potential migrants on their way to destination countries. The research seems built on the tragedy that took place in late 2013 when over 400 migrants died in two shipwrecks near the Italian island of Lampedusa. This research though indicates that Europe is the world's most dangerous destination for "irregular" migration, costing the lives of over 3,000 migrants this year alone. In the research, it was noted that since 2000 nearly 6,000 more migrant deaths occurred along the US-Mexico border and another 3,000 deaths from such diverse migration routes as Africa's Sahara Desert and the waters of the Indian

DOI: 10.1057/9781137451187.0005

Ocean. Many deaths occur in remote regions of the world and are never recorded. Limited opportunities for safe and regular migration drive would-be migrants into the hands of smugglers, feeding an unscrupulous trade that threatens the lives of desperate people.

To give examples of such fatalities, Malik (2011) writes that Dubai's skyscrapers are stained by the blood of migrant workers. In one of her trips to UAE, she discovered that a man – an Indian worker – had jumped from Burj Khalifa, the tallest building in the world. Gossip about the suicide was horrifyingly callous. "There is nothing remarkable about people being desensitized to suicides. London commuters on the underground can probably understand, but when the suicides are almost exclusively from one minority working in certain jobs, it is nothing short of inhumane. The dark underbelly of Dubai is never far away and sometimes we see the effect of this uglier side lying lifeless on a pavement."

The man, apparently an Indian cleaner who had been denied a holiday, was scraped off the floor on which he landed on and life went back to normal. The Indian consulate in Dubai has since revealed that at least two Indian expats commit suicide each week. The consul-general stated that most are blue-collar workers who are either semi-skilled or skilled. It seems to her that it is a place where the worst of Western capitalism and the worst of Gulf Arab racism meet in a horrible vortex. The most pervasive feeling is of a lack of compassion, where the commoditization of everything and the disdain for certain nationalities thickens the skin to the tragic plight of fellow human beings. Psychologically, these workers are isolated and alienated; practically, they are trapped by draconian sponsorship laws in the UAE, and in debt to agents back home. This is exacerbated by the fact that there is such little enforceable employment law in these markets. The above examples are particularly important in the context of migrant fatalities because on average 65 per cent of the total population in gulf countries is constituted by foreigners. However, these foreigners are not included in their domestic labour laws. In the Gulf Cooperation Council (GCC) – Bahrain, Kuwait, Oman, Qatar, Saudi Arabia, United Arab Emirates – the boom of labour migration began with the discovery of oil in the Gulf States in the 1970s. The boom triggered labour migration from South and Southeast Asia to meet the demand of manpower for construction and infrastructure development (Rahman, 2011; Jarallah, 2009, p. 6).

At the beginning of the age of labour migration, people used to travel short distances in order to seek work. People were less desperate to move

DOI: 10.1057/9781137451187.0005

than today. Therefore, migration was largely by choice. Today's migration patterns are extending further and growing faster than ever before. In 1990 migrants accounted for 15 per cent of the population of 52 countries (Council of Europe, 2000). The implication therefore is the less capacity for both origins and destinations to manage migration than the flow. Thus, in most cases, migration, especially unskilled or low skilled, has become an unsafe phenomenon. According to the IOM, while employment in developing countries tends to stagnate, there will be a further increase in their labour force, which is expected to rise from 2.4 billion in 2005 to 3 billion in 2020. On the other hand, the labour force in more developed countries is expected to remain at around 600 million by 2050. Therefore, developed countries' demand for migrant workers will increase significantly as a consequence of their ageing population (IOM, 2010). Unlike in the past where very few countries became involved in the labour sending–receiving process, now many countries particularly in South Asia, have started to participate in the export of labour. Though Asian labourers are scattered in many countries around the world, their main destination is the Middle East and Southeast Asia.

The growing flow is a result of growing demand. However, despite the growing demand and resultant response no efficient management and governance have been developed. In recent years there has been an increase in inter-country and inter-continental movements of migrants, with the majority moving from a lesser developed country to a developed one, instead of what traditionally has been a lateral movement from one poor country to another country with similar economic status. Still, international migration within developing countries is currently estimated to represent above 50 per cent of migrants (Susan, 2001). This "poor-rich dichotomy" in the migration realm has prompted receiving nations to overemphasize their own security issues rather than investing In migrants' safety and rights. This has crucial implications for the well-being of migrant populations. Receiving countries tend to forget that migrant populations are the response to their needs. Rather, they suffer xenophobia and alienate the migrants. Therefore, in most receiving countries, migrant populations are not included in the domestic labour laws or regulations. Thus, they are deprived of their basic human rights.

During the 21st century, migration from developing country to developed countries has increased exponentially, representing one-third of total international migration (Wickramasekara, 2006). Apart from

DOI: 10.1057/9781137451187.0005

the traditional immigration countries – the United States, Canada and Australia, Europe, the oil rich Middle Eastern states, and the rising economies of East and Southeast Asia, have become major destination countries as well. According to the Global Migration Report 2010, the number of international migrants in the Middle East is estimated to be 26.6 million (IOM, 2010), while 53 million are in Asia (Castles and Miller, 2009).

A study on 71 developing countries confirm that both international migration and remittances significantly reduce the level and severity of poverty in the developing world. It also reveals that on average, a 10 per cent increase in the share of international migrants in a country's population will lead to a 2.1 per cent reduction in the share of people living on less than $1.00 per person per day (UNFPA, 2006; Jr Adams and Page, 2005, p. 1645). Poverty is arguably the starting point, but other factors and practices may be involved in the process. A question that may arise, therefore, is, "is labour migration always voluntary?" as suggested by Siddiqui (2005, p. 1). The above analysis about the interplay between poverty and migration is significant when naïvely asked why the migrants do not choose to flee the abuses. In fact, migrants become trapped, in effect, by a number of factors (Ullah, 2010a), especially the female migrants who work as domestic helpers. For many migrants it is difficult to flee from these conditions because of the financial burdens that they incurred during their recruitment and travel. This dilemma can be called the "institutionalised dependency trap". The term implies the institutional and systemic nature of profit-making, the private and governmental mechanisms through which it is channelled, and the lack of oversight and legislative protection of domestic labour in the receiving countries (Ullah, 2010a).

Two forms of international mobility are identified: unskilled and semi-skilled workers on short-term contracts (overseas foreign workers), and skilled professionals and executives (Chia, 2006). It is evident that international migration flows are made up mostly of low-skilled workers. However, the relative weight of highly skilled immigrants has acquired a significant importance in recent years, unfortunately, because the major-ity of them move to or within the developed world. With regard to the type of work involved, the IOM highlights that the main occupations of both skilled and low-skilled migrant workers in developed countries are in the services sector. Most notorious abuses are perpetrated on the low and unskilled workers.

DOI: 10.1057/9781137451187.0005

Another significant feature of contemporary international migration that has attracted growing research significance over the past few years is the feminization of migration. Women now account for almost half of total migrants (IOM, 2008). Women have always taken part in migrations, but their presence varied depending on: (a) their origins, (b) the labour market situation both in the areas of origin and in the target areas, and (c) the migration policies in the immigration areas. Their presence has not always been visible, and they did not attract as much scholarly and political attention as they do today. How can one explain an increased interest in this area, when only a few years ago the subject was considered to be marginal? Migration studies have focused more on the geographic mobility of men, and did not consider women as equal protagonists. Could the so-called "feminization" of migration explain the interest in women and gender in migrations? In some groups men were primo-migrants but with the gradual feminization of migration, flows have reached a balanced sex ratio. Whereas women migrated first and became numerically predominant, one can now observe the opposite trend of "masculinization". Some authors rightly refer to this as "gender transition" whereas the term covers both trends. The visibility and growing interest for women in migration and, more recently, for a gender perspective in migration, is not only due to the changing migration patterns, and profile of migrants, but also the renewal of theoretical discourse in migration and gender studies; our focus on specific aspects of migration is triggered not only by genuine changes in migration trends, but also as a result of a long process of visibilization in the academic production on migration, women and gender (Mirjana, 2010). These trends may have existed earlier in migration history, but they were lost in the shadows of categories by those defining, recording, and analysing migration. Levels and severity of abuses thus are different from males to females. Females are considered more vulnerable than males overseas. Females are less resistant, a characteristic which perpetrators take as an advantage.

Table 1.2 demonstrates the migration status of the central players in terms of the workforce in Asia. There are a number of traditional labour-sending countries such as Bangladesh, the Philippines, Indonesia, China and Sri Lanka. The Middle East, Republic of Korea, Japan, Hong Kong SAR, and Singapore have now emerged as labour-receiving countries while India, Malaysia, Pakistan and Thailand are both labour-sending and labour-receiving countries. It is important to note that although

DOI: 10.1057/9781137451187.0005

TABLE 1.2   *Migration situations of Asian countries*

Country	Labour sending	Labour sending and receiving	Labour receiving
Bangladesh	√		
China	√		
Indonesia	√		
Philippines	√		
Nepal	√		
Sri Lanka	√		
Vietnam	√		
India		√	√
Malaysia		√	√
Pakistan		√	√
Thailand		√	√
Middle East			√
Brunei Darussalam			√
Taiwan			√
Japan			√
Republic of Korea			√
Hong Kong SAR			√
Singapore			√

Source: Adapted from Piyasiri (2002).

labour-sending countries are mainly economically impoverished, they have surplus employment. While on the other hand labour-receiving countries have shortages in manpower. Without adequate indigenous manpower developed, as well as middle-income countries, must receive workforce from countries with a surplus in labour supply.

Recent trends of labour migration in South Asia

Southeast Asian contemporary labour migration developed in the 1960s and expanded during the 1970s and 1980s. Following the oil shock of the 1970s the oil-rich Gulf countries underwent major construction and development from projects that drew considerable numbers of migrants from Asia and Southeast Asia. Thailand, the Philippines, and Indonesia saw this opportunity as an economic lifeline, and started overseas labour deployment programmes to benefit from the employment opportunities in the Middle East.

During the same period of development, Singapore and Malaysia also experienced labour shortages; they became attractive destinations for a regional, temporary labour movement. The growing political and

DOI: 10.1057/9781137451187.0005

economic interconnectedness within and between states in the region, particularly after the formation of the Association of Southeast Asian States (ASEAS), also promoted individual personal mobility. As economic and demographic disparities between states became more obvious, the local geographical and economical contexts assumed greater importance in shaping and facilitating cross-border labour flows in southern Asia. In 2009, Asian countries received US$162.5 billion in global remittances, which is 39 per cent of the global total.

Asian countries have a surplus of labour. Thus, Asia has been one of the top senders of labour migrants across the world; China, Bangladesh and India are among the top ten emigration countries worldwide. According to the 2008 World Migration Report "Asia is characterized by being the largest source of temporary contractual migrant workers world-wide as well as by the large intra-regional flows of migrant workers" (IOM, 2008). Moreover, according to the World Bank, Asian countries make up four of the top ten migration corridors worldwide: Bangladesh–India (3.5 million migrants in 2005), India–United Arab Emirates (2.2 million), the Philippines–US and Afghanistan–Iran (both 1.6 million) (World Bank, 2008).

Labourers migrating from Asia are mainly unskilled; they are frequently engaged in low-paid jobs, such as construction and infrastructure projects, maintenance activities, and the service sectors, especially domestic services (Piyasiri, 2002). Recently, manpower migration from Asian countries has shifted to the Middle East, where labour markets depend on the workforce of South Asian countries. The IOM reports that "approximately 37 percent move to OECD countries; of the remainder 43 percent migrate within the region and the rest migrate to other countries outside the region" (IOM, 2010).

Table 1.3 shows documented labour flows for a number of manpower-sending countries. The Philippines and Pakistan began exporting their workforce in 1972. Thai labour emigration started the following year. In 1976 India, Indonesia, Bangladesh and Sri Lanka joined this list as a labour exporting country. But the Philippines have retained its first position from the beginning. In 1998 the Philippines sent 562400 workers all over the world while Indonesia was the second largest labour-sending country in 1998 with 411600 workers deployed. In the years 1972–1982 the Philippines became the largest labour exporting Asian country; Pakistan and India were the second and third largest exporters, respectively. In all Asian workforce-sending countries, the

DOI: 10.1057/9781137451187.0005

TABLE 1.3   *Gender-wise distribution of Bangladeshi migrants (1976–2014)**

Country	Male	Female
Saudi Arabia	2,631,566	32,126
UAE	2,310,492	86,784
Kuwait	480,108	7,659
Oman	878,582	22,317
Bahrain	280,420	6,656
Qatar	290,864	3,697
Malaysia	707,339	6,521
Singapore	503,314	868
Libya	119,676	531
Jordan	71,531	60,953
Lebanon	101,886	87,529
South Korea	30,368	–
Italy	55,162	463
Brunei	42,377	82

Note: *Until September.
Source: Adapted from Bangladesh Manpower, Export and Training (BMET, June 2014).

migrant worker outflow tends to increase every year, however from 1985–1991 Indian migrant outflow actually decreased. The mid-1990s saw the highest sustained outflow of labour from Asian source countries. Bangladesh is one of the top senders of labour migrants in Asia. Factors driving their manpower migration include stressed demographics, and the economic inequality and social consequences of globalization (Khondker and Rahman, 2009). Bangladeshi migrants are both skilled and unskilled, and their main destinations are the Middle East, and East and Southeast Asian countries.

Bangladesh's era of labour export began in 1976. Initially, few countries received manpower from Bangladesh. Today Bangladeshi migrants work in more than 20 countries across the world. The migration history of Bangladesh can be divided into two periods: the pre-independence period (before 1971), and the post-independence period. During the pre-independence period, those seeking permanent settlement dominated migration. During that time large numbers of Bangladeshi, mainly Hindus, migrated to India because of war and persecution. In the early 20th century many inhabitants of the region moved to Western Europe, North America, and Oceania in the search of permanent settlement to contribute to the evolution of the Bengali community overseas (World Bank, 1981; Gardner, 1995; Knights, 1996).

DOI: 10.1057/9781137451187.0005

Post-independence migration was dominated by temporary migration of labour, although a large number of Bangladeshis from higher and middle-income groups continued to migrate for permanent settlement in developed countries and unskilled labourers sought out livelihood rather than social mobility. During the early part of this period the destination of choice for emigrating Bangladeshis was mainly the oil-rich Gulf countries. However, this trend declined dramatically in the late 1980s (Arnold and Shah, 1986). The Gulf War in 1990–1991, and the 2003 US invasion of Iraq, caused serious disruption to the regional economies and suppressed demand for foreign labour.

The economic growth miracle of Southeast and eastern Asia served as a magnet for unskilled and semi-skilled labourers seeking temporary employment. Bangladeshi migrants flocked to this region as well. Singapore, Malaysia, South Korea, Brunei, Hong Kong and Thailand became the major labour-receiving countries. Japan also continued to host large numbers of documented and undocumented migrants from Bangladesh (Khondker and Rahman, 2009). This is obviously a promising picture. The above analysis offers an important perspective, showing that governments of those countries are concerned about the safety and security of their citizens while the migrant community at large remains outside of their safety net. Therefore, the position of migrants in receiving countries becomes more vulnerable than ever. It is important that the respective governments make their citizens aware of the role that migrants play in their countries' development. This will encourage individual recruiters to treat them humanely.

While there is evidence that the Bangladeshi migrants are repatriating, there is another trend emerging which is the increase in the number of female migrants. The countries that are sending back male Bangladeshi migrants have started to recruit female labour. However, the female migration is about 2 per cent of the male counterpart. This is one of the reasons why this research has not highlighted female migration from Bangladesh. Employers in Hong Kong have recently shown interest in recruiting female workers from Bangladesh. As a test case, last year in 2013 a small number of female migrants were recruited in Hong Kong as domestic helpers. However, most of them returned or were repatriated within a few months because they failed to adjust to the new environment. Job growth in Eastern Europe has opened up new opportunities for women in sectors such as garment manufacturing, caregiver industries and electronics. This has led to an

DOI: 10.1057/9781137451187.0005

increase in the number of women from Bangladesh moving overseas to find employment (IOM, 2010).

Occupational hazards and fatalities correlates

Migrant fatalities are associated with occupational hazards, so minimizing their risk of becoming a victim is important. This requires respective governments to be proactive. Migrant workers labour in all seasons and weather conditions, including extreme heat, cold, rain and bright sun. Work often requires stoop labour, working with soil and/or heavy machinery, climbing and carrying burdensome loads; all of which may lead to serious physical health problems. Although "MSFWs (migrant and seasonal farmworkers) and their families suffer from the same health problems found in the general population, the occupational hazards, poverty, substandard living conditions (Hansen and Donohoe, 2003, p. 1)", and language and cultural barriers that they face result in health hazards unique to migrants. As a result, the average life expectancy of MSFWs is a mere 49 years, compared with the national average of 75 years.

Infectious disease

Migrant workers are at increased risk of contracting a variety of viral, bacterial, fungal and parasitic infections. Migrant workers who are involved in 3D category jobs are approximately six times more likely to have tuberculosis than the general population; up to 44 per cent of migrants have the positive purified protein derivative of tuberculin skin tests (Hansen and Donohoe, 2003).

Chemical and pesticide-related illnesses

The full extent of acute and chronic pesticide poisoning among MSFWs is not known due to the lack of formal reporting systems, the reluctance of workers to report poisonings, workers' inability to seek medical treatment when accidents occur, and a dearth of physician knowledge and training in recognizing and treating pesticide-related illnesses (National Rural Health Care Association, 1986; Hansen and Donohoe, 2003).

DOI: 10.1057/9781137451187.0005

Reproductive health

Prolonged standing and bending, over exertion, dehydration, poor nutrition and pesticide or chemical exposure contribute to an increased risk of spontaneous abortion, premature delivery, foetal malformation and growth retardation, and abnormal postnatal development. Migrant workers are exposed to a wide range of pesticides, solvents, oils, fumes, ultraviolet radiation from chronic sun exposure, and biologic agents such as human and animal viruses (Hansen and Donohoe, 2003).

Social and mental health

Migrant workers face numerous sources of stress, including job uncertainty, poverty, social and geographic isolation, intense time pressures, poor housing conditions, intergenerational conflicts, separation from family, lack of recreation, and health and safety concerns (Hansen and Donohoe, 2003).

We can easily understand the level of vulnerability to health problems that migrant workers face, and it is especially important to recognize the human brutality perpetrated towards them. We are reminded of Samalatha, a Sri Lankan citizen, whose dead body arrived from Saudi Arabia with most of her body parts missing. In 2013, another Sri Lankan domestic maid in Saudi Arabia returned with 24 nails in her body. Few incidents rarely come to light, and when such incidents lead to death, various ways are adopted to conceal the real cause behind the death. In the cases mentioned above, there were no reports that the perpetrators were brought to justice. This kind of practice of impunity encourages the perpetrators.

The above points made by Hansen and Donohoe, and Baron have crucial implications for migrant fatalities. Severely compromised social, mental health and reproductive health, and exposure to infectious disease, may eventually lead to serious health risks if not given medical attention. In recent years incidences of abuse of female South and Southeast Asian domestic workers in the Middle East have surfaced. Reports and studies conducted among migrant workers in the Middle East have revealed substantial forms of labour exploitation and abuse, from non/underpayment and long working hours to verbal and physical abuse (e.g., Afsar, 2009, 2011; Ullah, 2013; Godfrey, Ruhs, Shah and Smith, 2004; Human Rights Watch, 2007, 2010a, 2010b; Jureidini, 2004).

DOI: 10.1057/9781137451187.0005

Nonetheless, the harms which female domestic workers experience still remain largely unrecognized since they usually work as live-in house-maids in their employer's homes, isolated from the public (Godfrey et al., 2004).

Institutionalized dependency trap

It is important to touch upon the Institutionalized Dependency Trap (IDT) in order to understand why and how migrant casualties occur, and if such incidents could have been avoided or minimized. IDT in fact is applicable to cases in the Middle East and Gulf countries. Migrants are vulnerable to abusive and exploitative conditions in Middle Eastern countries, especially for those who live within their employer's household (Jureidini, 2004, p. 67). Housemaids in particular, have become victims of humiliation, violence, long working hours, and delayed salaries (e.g., Jarallah, 2009, p. 7). We may seem to attach a lot of focus on the Middle Eastern countries. This is because reports of the most frequent abuse come from this region. Of course, not every woman who goes abroad to a Middle Eastern country as a live-in domestic worker experiences abuse, or else this would mean every Arab employer is exploitative and abusive. Jureidini (2004) reports that many foreign workers in Lebanon testified to receiving "decent respectful treatment and have grown to like their employers" (p. 70). However, the policies in place, the *kafala* system for example, lead most employers to be abusive and exploitative.

The abuse a female migrant might experience may be a result of the IDT, which can arise due to a combination of factors making it difficult for her to escape exploitative conditions. Their vulnerability to labour exploitation, and abusive labour conditions, starts in the country of origin, and arguably is an effect of poverty. For most poor women and their families, especially for those from rural areas, going abroad is a huge financial investment and means going into debt or selling assets in order to finance the necessary expenses that arise prior to the journey. This indebtedness, together with a confluence of other factors can enhance a woman's vulnerability at different stages of the migration process. In the end, she could find herself trapped with the need to pay back loans while supporting her family back home.

Furthermore, the controls of the employer and the weak or non-existent legislative and protective structures in the country of

DOI: 10.1057/9781137451187.0005

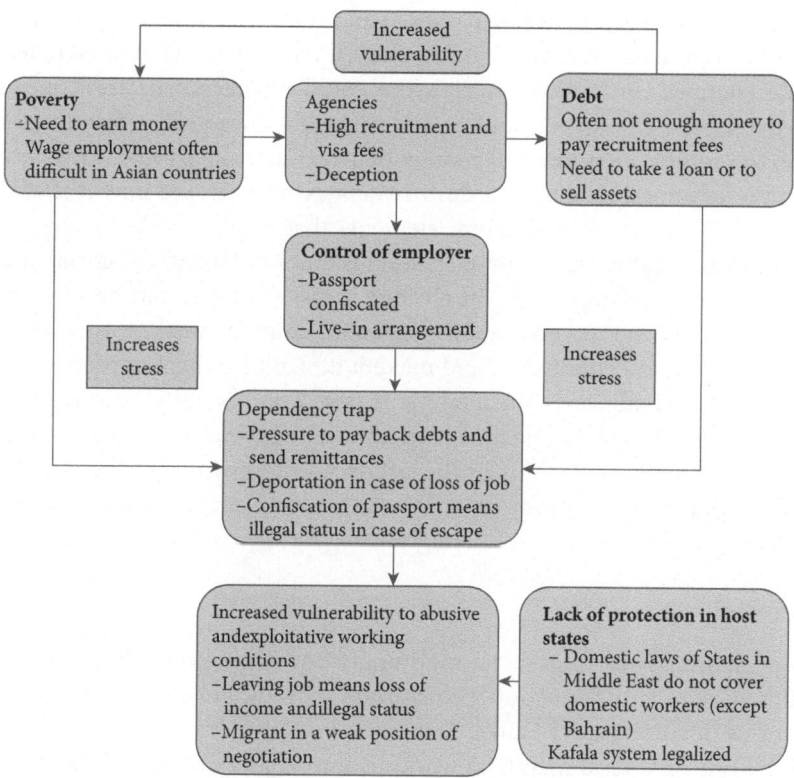

FIGURE 1.3 *Dependency trap*

Source: Authors.

employment make it difficult for them to escape their workplace and put them in a weaker position from which to negotiate their conditions. The dependency trap can be described as "institutionalized" because of its systemic nature and the legislative flaws which contribute to it.

It is relevant to expand a bit on the effects of the IDT. Unaware of the position they are getting themselves into, the migrants become victims of IDT when they realize that the working environment is becoming or has already become unsafe and they are unable to extricate themselves from the situation. One of our authors interviewed a Bangladeshi worker in Saudi Arabia. He explained how he was often brutally beaten by his employer. He could not forget the most gruesome experience he had in 2012 when he had to talk to the employer's wife on a matter related to his

DOI: 10.1057/9781137451187.0005

work. The employer became paranoid and started beating him mercilessly and it continued until he lost his sense. A few days later he decided to leave the country. He however realized he could not terminate the contract, whereas the employer could at any time. If he wanted to terminate he had to return an amount of money he never actually earned. This is one of the examples of abuses at destinations. Now let us see how it is at the origin. There are four different elements that play key roles in the IDT: (1) poverty; (2) recruitment agencies; (3) employers; and (4) government policies of receiving states. In all four areas, attributes can be identified which enhance the vulnerability of female migrant workers to exploitative and abusive conditions, and make it difficult for them to escape those very same conditions. The analysis is based on several case studies and reports conducted among labour migrants, in general, or female domestic migrants, in particular, during the last decade. It provides a general picture of the practices and problems that arise for women from Southeast Asian countries who use agencies to migrate to countries in the Middle East.

Poverty

For many labour migrants emigrating from developing countries seeking work in another country constitutes a livelihood strategy, and a way out of poverty. In Sri Lanka, for example, it is common for families to have one or several members working abroad as a means to extend their income base and even support large relative groups. Families with relatives working abroad are less likely to be poor (Shaw, 2010, p. 66). Solely considering the economic reasons for female migration probably generates a too simplified picture. Motivations are frequently multi-faceted, with more freedom and less societal constraints being but two other possible explanations for the decision to work abroad. However, poverty and the need to generate income are considered as the main factors.

High indebtedness increases vulnerability to poverty. Instead of stepping out of poverty, the indigent become trapped in a "cycle of poverty". This phenomenon occurs when poverty becomes a generational burden from which families cannot escape. Due to a lack of education and the resources necessary to invest in social and economic mobility, affected people rarely manage to break this cycle (Marger, 2007). Payne (2005) explains that besides generational poverty, a specific event may trigger a fall into a cycle of poverty. Not being able to pay back loans can certainly be one factor. If loans go into default, women face crushingly high interest rates. Debt decreases their chance of breaking the cycle of poverty,

DOI: 10.1057/9781137451187.0005

and very likely sinks them further into it. The study by Afsar (2009) revealed that limited options to finance migration, and the high interest rates of moneylenders, are two major constraints for labour migrants (p. 11). The fact is that many workers, who are already poor, are under immense pressure to pay back loans. This significantly inhibits their ability to leave their job if they face abuse or other difficulties. Human Rights Watch (2007) concludes that high "fees place workers at risk of later exploitation by employers, because they feel trapped in abusive employment situations as a result of the debts they must repay to labour agents, subagents, banks, and moneylenders" (p. 22).

Recruitment agencies

It is important to understand the involvement dynamics of recruiting agencies in the recruiting process because there are widespread claims that exploitation begins from that point. One important way for a worker to find employment is through personal connections (Ullah, 2010a; Afsar, 2009). Thereby, the prospective migrant purchases a visa through relatives, friends, or other acquaintances already working abroad (Malecki and Ewers, 2007; Al-Najjar, 2004, p. 29). Some other sending countries have established governmental agencies which organize, regulate and monitor the manpower migration sector. Examples are the Sri Lankan Bureau of Foreign Employment (SLBFE, 2009) and the Bangladeshi Ministry of Expatriates' Welfare and Overseas Employment. In sending and receiving countries, private agencies too are mushrooming in order to facilitate migration.

From their home country a migrant worker can apply through an agency for work abroad. Potential migrants receive applications from workers via unlicensed subagents who operate in remote villages. The subagents, or agencies, then make the necessary bureaucratic arrangements and connect the workers to their prospective employers (Ullah, 2010a; Afsar, 2009; Human Rights Watch, 2007). Alternatively, a prospective employer can hire a worker via an agency. In the receiving countries of the Gulf States, most recruitment agencies are privately hired by employers. In Bahrain, the employer pays a fee to the agency, including a flight ticket, and selects a worker. The employer also obtains a work permit for the prospective employee (Al-Najjar, 2004). Research has revealed many illegal and fraudulent practices by private agencies, which arguably increases the risk of workers ending up in a vulnerable situation. Human Rights Watch (2007)

DOI: 10.1057/9781137451187.0005

reports that agencies often charge higher fees than allowed by domestic laws in sending states. Overcharging, or double charging, is common practice among some agencies and subagents (Afsar, 2009, p. 27).

The lack or falsification of documents is yet another factor contributing to the vulnerability of migrant workers. Studies have reported that contracts were not provided to, or signed by, the workers (Afsar, 2009; Sabban, 2004), that if the contract was shown to the workers it was not in their native language, or the content was not explained to them (Jureidini, 2004, p. 81), and that often a contract was shown but it was an embellished one prepared locally by the agents, meaning that it was fake. Al-Najjar (2004) found that among 34 migrants from Sri Lanka, the Philippines and Indonesia about 44 per cent had signed a contract prior to their arrival in the destination country. On some occasions the agents themselves signed the passport on behalf of their illiterate clients (Siddiqui, 2005). All of which can lead to dire troubles for the workers. In the absence of a contract, workers are in a weak position to demand their rights. A false signature on the passport can result in conflict with immigration authorities.

Finally, a serious problem that Frantz (2008) describes is that regarding workers who arrive in their country of work, but never obtain the promised entry they paid for (p. 623). In the Gulf States, foreign workers stay usually on a temporary basis as guest workers. For this reason they need a visa signed by a local sponsor, usually the employer, in order to be allowed to enter and stay. Without such a visa, a foreign worker is officially illegal (Frantz, 2008, p. 623; see also below "Government Practices"). Irregular and illegal practices are a general problem for labour migrants. Human Rights Watch revealed similar problems in the case of male foreign construction workers in the United Arab Emirates (Human Rights Watch, 2009, pp. 2–3) and Kuwait (Human Rights Watch, 2010b, p. 24). Workers may be forced to sign contracts in Arabic upon arrival (Human Rights Watch, 2009, p. 2). This implies that a straining relationship between employers and employees begins as soon as the workers arrive, and is a contributing factor in the ensuing fatalities.

Employers

The control that employers may exert over their domestic employee is another factor that advances the dependency trap. As stated above, a foreign worker needs a sponsor in order to be able to enter and work

DOI: 10.1057/9781137451187.0005

in an Arab country (Frantz, 2008, p. 619; Sabban, 2004, p. 100). Under this system, called *kafala*, employees are forbidden to change employment without their employer's consent, though this is illegal by law it is a common practice (e.g., Siddiqui, 2005, p. 17). The system is meant to make sure that the incoming worker is employed and then repatriated after the end of a contract (Godfrey et al., 2004). In fact, *kafala* in effect ensures that employees are enslaved (Rahman, 2011).

According to Frantz (2008), the high recruitment costs for an employee, costs which are non-refundable in the case of escape, often means that employers develop a sense of ownership over their employees (p. 628), therefore they can do whatever they want with them and to them. Qama Rahim, the Former Secretary General of South Asian Association for Regional Cooperation rightly said that "workers become virtual slaves when their passports are taken away by employers, as their status in a foreign land becomes illegal without passports" (Chowdhury, 2008). Moreover, problems can arise for the employee if the employer lets their visa expire to avoid paying the fees for its renewal.

Maids often live within their employer's household. A positive aspect is certainly that the domestics can save money, which they would otherwise have to spend on rent for accommodation. This arrangement, however, makes them more vulnerable to abuse and exploitation (Human Rights Watch, 2007, p. 4), and a feeling of isolation is a common consequence (Sabban, 2004, p. 92). Al-Najjar (2004) explains that the socio-economic conditions of domestic female workers are such that they refrain from insisting on clear contracts with their sponsors. Consequently, after arrival they often discover that working conditions are not what they expected or what they were told (pp. 30–31). Non-payment or delays in payment leave the worker in a precarious situation, especially if she had incurred debts prior to departure. Furthermore, the study by Godfrey et al. (2004) in Kuwait showed that a major problem indicated by many workers, both male and female, was physical and verbal abuse (p. 54). Similar findings regarding the abuse of domestic workers are provided by Sabban (2004) in the case of the United Arab Emirates (p. 97).

Government practices of receiving countries

The dubious practices of private agencies and subagents, and the *kafala* system are but two elements that portray the institutionalized character of the dependency trap. Immigration laws, and policies, in

DOI: 10.1057/9781137451187.0005

many labour-receiving countries act to further institutionalize the dependency trap. This relationship between the state and the labourer will be further explored. The vast legal gap between sending and receiving countries renders foreign domestic workers unprotected in receiving countries. With the exception of Bahrain (Jarallah, 2009, p. 10), Middle Eastern countries designate domestic work outside the jurisdiction of their labour laws (ILO, 2011b; Sabban, 2004, p. 99). The Jordanian Government's recent effort to introduce standardized working contracts, which forbid the withholding of salaries and confiscation of passports, do not include domestic workers (Frantz, 2008). Similarly, Kuwait's 2010 labour law that grants extended rights to private sector employees does not cover domestic workers (MOSAL, 2010, Art. 2).

The exclusion of domestic workers from laws regulating the workplace, coupled with restrictions on their ability to associate, not only limits their workers' rights but also greatly increases their risk of abuse. The Committee on Economic, Social and Cultural Rights (2004) expressed its concern regarding female domestic workers in Kuwait due to the fact that the National Labour Code failed to protect them, describing the conditions as "not dissimilar to forced labour" (p. 17).

It is without doubt that the patriarchal cultural and legal environment of the Gulf States' amplifies the negative effects of the *kafala* system. In Kuwait workers who have left their employer without permission are usually arrested and deported (Human Rights Watch, 2010a). They also face a fine of up to KD 600 (US$2,077), or imprisonment for up to six months prior to deportation (Human Rights Watch, 2010b, p. 35). Such policies actively "inhibit workers from reporting abuse" (pp. 2–3). In the United Arab Emirates, employer confiscation of passports is actually encouraged by the requirement that workers' visas are to be cancelled and their passports turned in if they leave their employers' service (Human Rights Watch, 2009). The recent legal reforms in Bahrain and Kuwait in 2009, which relax the requirement that employees need the consent of their employer to change jobs, do not cover domestic workers (US Department of State, 2010a).

Some efforts are claimed to have been made by the Arab States, including the establishment of a Bahraini central authority to control the *kafala* system (Shaham, 2008, p. 8) and legal sentencing of abusive employers (US Department of State, 2010a). As examples of their own efforts to address these legal failings, in early 2014, Qatar abolished the *kafala* system and ordered the creation of shelters for runaway domestic helpers. However, all of these efforts are insignificant in the face of such an enormous and long-standing need. Some say this is a mere response to the call of the world to treat migrant workers humanely. To some extent the new policies and initiatives are not even successfully implemented. Regulations to supervise employment agencies in Lebanon (Jureidini, 2004, p. 64), and a blacklisting system in Saudi Arabia, fell short of their aim (US Department of State, 2010d, p. 47). As long as domestic workers are not covered by the receiving countries' laws, and are legally tied to their employers, they face exploitation and abuse.

In the event of abusive and exploitative working conditions, it is difficult for a foreign worker to leave his employment. On the one hand, they most probably have incurred huge financial debts in order to finance migration in the first place. On the other, the pressure of paying back loans, and remitting money to the family, makes quitting the job financially expensive. The worker has virtually no legal rights. By signing a contract, a foreign worker is legally bound to be with the employer. The employer can use outstanding wages, as well as the possession of their passport, as a means to subjugate the worker. If a worker still decides to escape, he/she becomes an illegal migrant. In most Middle Eastern countries, quitting a job constitutes a breach of contract, the consequence of which is deportation. The extent to which this trap is institutionalized becomes apparent not only when looking at the laws of the receiving countries, but also at private recruitment agencies, which often deceive and overcharge their clients.

The labour flow of women from South and Southeast Asian countries emigrating as housemaids in the Middle East is well established. In light of the numerous reports of the seemingly systematic abuse of migrant workers in general, and housemaids in particular, it becomes imperative that the global community finds ways through which domestic workers can be better protected. As has been argued, the dependency trap often starts with the factor of poverty itself, which is difficult to eliminate in the short term. It is however possible that the channel of labour migration becomes safer for migrant workers. This means, first of all, that migrant workers and their families need not finance the trip through the use of exploitative loans at high interest rates. One possibility is that sending-state governments provide funds or small credits, at affordable

DOI: 10.1057/9781137451187.0005

interest rates, to finance the migration process. Moreover, both sending and receiving states have to develop and improve mechanisms to fight the illegal practices of recruitment agencies. In the receiving countries, domestic work has to be included into national labour laws, and institutions have to be established not only to increase the protection of foreign domestic workers, but also to enable them to seek help in the event of abuse. We have devoted quite a large part of this book to talking about the trap. We also argue that those who are trapped face more brutal treatment from employers than those who are not.

No systematic study is available on how many migrant workers die and suffer fatalities abroad and why. No specific policy is in place to deal with the cases of death of migrant workers, and their money that remains in the destination countries. This book tries to address those issues.

Significance, objectives and methods

In the last 5 years Bangladesh, whose economy relies heavily upon remittances, received far too many repatriated bodies of migrant workers (Daily Star, 2008). Official causes of the deaths have sited job insecurity, tension at work, poor working conditions, and unhealthy food habits. The debate over the events of these deaths has at least secured major policy concessions for Bangladeshi expatriates from Saudi Arabian authorities, but in actuality very little has been done. Around 44 per cent of the Bangladeshi migrant workers who returned home in coffins, between 1 January and 9 May in 2009, had died of cardiac arrests in West Asian and Southeast Asian countries, raising serious questions about the living standards of migrant workers in those countries. Cardiologists generally claimed that acute tension caused by uncertainties of income and unhealthy food habits may have led to the deaths by heart attacks, while labour rights activists are emphatic that mental tension caused by low income, debts, and lack of medical care led to such deaths. A total of 904 bodies of migrant workers were repatriated from different countries between January and May; 391 of which died of cardiac arrests, 268 in workplace accidents, 62 in road accidents, 115 of other illnesses, and the remaining due to various other reasons. Among the 391 deaths by cardiac arrests, 119 were in Saudi Arabia, 82 in Malaysia, 72 in the United Arab Emirates, 35 in Kuwait, 16 in Oman, ten in Qatar, ten in Bahrain, seven

DOI: 10.1057/9781137451187.0005

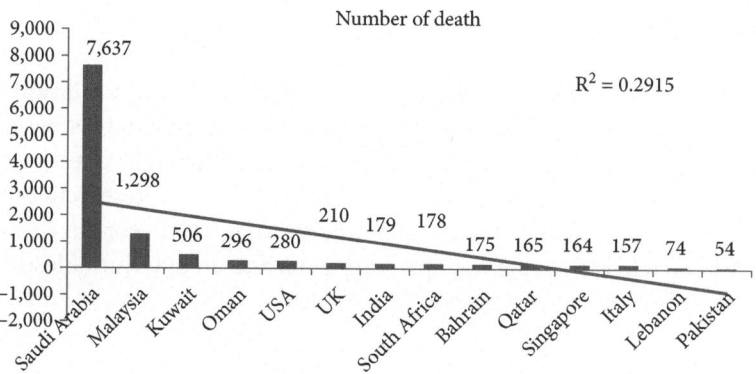

FIGURE 1.4   *Dead migrants from selected countries (2009–April 2012)*
Source: Adapted from BMET (2012).

in Singapore and two in Lebanon – the major destinations abroad for Bangladeshi labourers (BMET, 2014).

The number of deaths of migrant workers abroad has been increasing exponentially since 2004. Last year, the number of dead bodies transported home was 2,237; far more than the annual averages from previous years, which was 1,673 in 2007, 1,402 in 2006, 1,248 in 2005, and 788 in 2004. It is usual that our workers have acute mental tension, as they work far from their relatives, and quite often their incomes are not up to their expectations. "In the Middle Eastern countries the migrant workers also eat more meat and other fatty foods than vegetables", such food habits increase the risk of heart attacks. It is surprising that many deceased workers aged between 25 and 40 had been medically fit at the time of leaving home, but died within a few months of getting to their workplaces abroad (Ullah, 2014; see also 2008; Daily Star, 2008; Bangladesh Bank, 2007).

While some studies do exist on the dynamics of usages of remittances, none exists on how the remittances are used after the death of the migrant and what general procedure is in place for recovering the money left in the destination country. Therefore, this research is significant in a number of ways, and for a number of factors. To our knowledge this is a new research, the only one of its kind in the field of migration and remittances. This study will significantly add to the scholarship on the financial behaviour of remittance receivers.

DOI: 10.1057/9781137451187.0005

Methods

In line with the objectives discussed for the present study, relevant data have been gathered from numerous sources. The research is based on both secondary and primary data collected through questionnaire surveys, case studies, and key informant interviews. Researchers have employed both qualitative and quantitative analyses. Secondary sources have been used to supplement primary data, and analyse variables within the broader context of migration dynamics.

Data collection

Data were collected in both the origin and destination countries. This was necessary to achieve the objectives of the study, however in the absence of a sample framework data collection was a serious challenge. The differential estimates between the government agencies and media reports served as a serious predicament to the process of estimating a sound sample size. Data were collected between April 2010 and January 2013.

This study used both structured and unstructured questionnaires. We deliberately used Bangla in formulating the questionnaires; we recruited research assistants locally so that participants were not inhibited in understanding the questions. They were explained the purpose of the research, and assured that they could withdraw at any time. If they were uncomfortable with any particular question they were at liberty not to answer that question. Research assistants were given appropriate training before they began collecting data. The questionnaire covered information related to socio-demography, health, torture, compensation, and government assistance. In order to examine the nature and degree of vulnerability, the secondary data have also been instrumental for this study. In most cases, we had to count on the media for ascertaining the number of migrant casualties that arrived from various destinations to Bangladesh, but we did not limit our review to only media sources. Extraordinary support was received from government agencies such as the BMET office in Dhaka. The ethical approval was obtained from the IRB, the American University in Cairo (AUC) before we began our research.

DOI: 10.1057/9781137451187.0005

TABLE 1.4   *Distribution of samples (origin and destination country)*

Name of districts	f
Chapai Nawabganj	26
Bogra	18
Chuadanga	10
Serajganj	06
Joypurhat	05
Jessore	03
Natore	03
Rajshahi	02
Meherpur	02
Sathkhira	01
Total	**76**
Distribution of samples (destination country)	
Saudi Arabia	6
Kuwait	4
Malaysia	11
Qatar	2
Total	23

Source: Field data (2014).

Selection of sample

Due to the nature of the sample distribution, we had to utilize a mixed sampling technique in Bangladesh. In some cases, we resorted to snow-balling to trace respondents; in the rest we were able to select respondents through stratified random sampling. Obviously, in destination countries, we had to use snowballing as our primary method. Interviews took place in locations where respondents were most comfortable. Socio-economic status, and other relevant data, were collected from the household head (HH) of each family unit, usually the father; in the event that they were not available, the second HH, that is, the mother, would participate.

Government officials/stakeholders

Data processing and analysis

A data entry operator was recruited after the questionnaire survey was completed. Field data were cleaned and entered into the SPSS program.

DOI: 10.1057/9781137451187.0005

Both descriptive and quantitative approaches (inferential statistics) were employed in analysing the primary data. Descriptive statistics tend to provide a general and simplified picture of particular issues. This study encompasses graphical, tabular and textual representation of the data. Frequency tables, mean, median, mode, standard deviation were prepared to show the demographic and socio-economic profile of migrant casualties and their families.

Inferential statistics were used in this study, as they are more inter-pretive than descriptive statistics. This study employed cross-tabulation and correlation. The Chi-square test was applied to show the association between arrays of variables. This study also applies a multiples regression model in analysing the determinants of remittance sent by the labour migrants. Thus step-wise regression was employed to determine the variables closely associated with remittance flow. In this connection it should be noted that studies on dead Bangladeshi migrants is relatively scant despite the fact that the phenomenon is recurrent and affects a large number of rural families. This is the rational purpose for conduct-ing our research.

In terms of validity and reliability, we remained careful about any kind of bias and cross-checked the data every day to avoid any redun-dancy and duplication. The following map shows the study area in Bangladesh.

As regards the organization of chapters, the second chapter analyses the central issue this volume deals with, which is the right of migrant workers. Migrant populations, migrant workers in particular, are extremely vulnerable to human rights violation, racism, xenophobia and discrimination. Most receiving countries' labour laws exclude migrant workers. This in fact has closed the option for migrant workers to seek redress in the event that they fall victim to any kind of human rights violation. We argue that the lack of protection encourages employers to be exploitative and reluctant to give workers the attention they deserve. The Convention on Migrant Workers has forged new ground, and placed human rights in the specific context of migrant rights. As its salient feature, the Convention protects all migrant workers and members of their families, irrespective of their legal status.

Chapter 3 analyses the profiles of the migrants who died, and those of their families residing in Bangladesh. We attempted to connect their family morbidity, demographic and socio-economic histories in order to further uncover the level of vulnerability that they are subject to after

DOI: 10.1057/9781137451187.0005

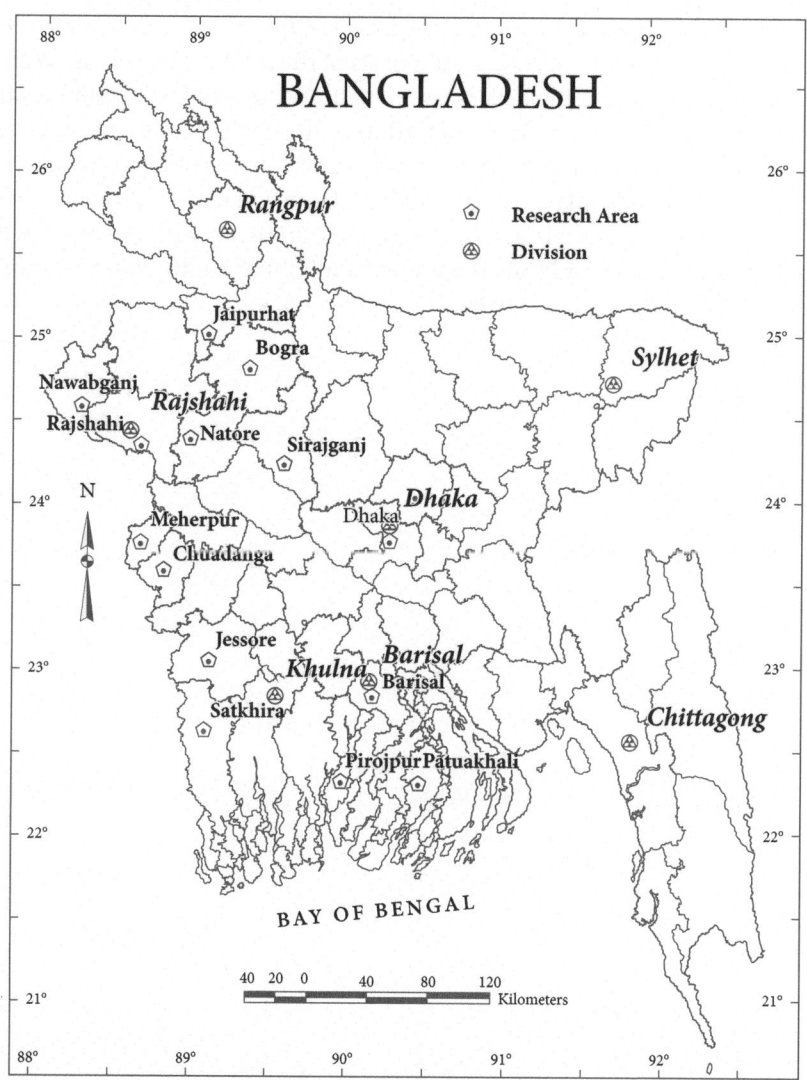

MAP 1.1   *Study areas*
Source: Digitized by authors.

the death of a loved one. This chapter also sheds light on a number of significant facts, such as how much money was left in the destination country, what mechanisms were used to recover the money, and how the families were compensated by the government or employers, if any.

DOI: 10.1057/9781137451187.0005

Chapter 4 attempts to charter the range of migrants' vulnerabilities to existing migration policies at both points of the migration process. While migration is structurally necessary to many of the growing economies in Asia, existing legislation does not address the challenges migrants face. A common ground among the sending and receiving countries' policies remains tremendously limited. The last chapter makes concrete recommendations to close this gap. We hope that these recommendations will help better understand how the issues of migrants' rights could be more humanely and progressively handled.

DOI: 10.1057/9781137451187.0005

2

Migrants' Rights and Gaps in Protection

Abstract: This chapter attaches emphasis on the central theme of this volume that deals with the rights of migrant workers. Having offered the overview of migration and the background of the research, this chapter analyses two key issues: the current protection mechanisms for migrant workers and their remittances. The chapter also explores the issues of remittances that support the livelihood of the migrants' families, and the welfare fund for the migrants while they are trapped in the destination countries. The authors argue that the lack of protection encourages employers to be exploitative and reluctant to give workers the attention they deserve. The case study examples from KSA, Kuwait and Malaysia are presented to illustrate the severity of miseries of the migrants and the tale of stories of the dead which the families of the bereaved mistrust.

Ullah, AKM Ahsan, Mallik Akram Hossain and Kazi Maruful Islam. *Migration and Worker Fatalities Abroad.* Basingstoke: Palgrave Macmillan, 2015. DOI: 10.1057/9781137451187.0006.

The previous chapter offered an overview of migration as a whole and the background of the research on which this manuscript is built. This chapter analyses two issues: the existing protection mechanisms for migrant workers and their remittances. This is important particularly for understanding the scope of the vulnerability of migrant populations while employed in foreign lands. The chapter also focuses on the remittances that contribute to the livelihood of the migrants' families, and the welfare fund that aids migrants while they are in trouble in the destination countries.

The level of protection that migrant workers have depends largely on the policies in place in both origin and destination countries. The level of vulnerability also depends on the type of labour, that is, skilled, unskilled or semi-skilled, in which the worker is employed. For example, the vulnerabilities that persist in the Middle East vary widely with those in Southeast Asia. The lack of protection of the foreign domestic workers in the Middle East is widespread. This chapter focuses on how the lack of policies to protect migrant workers originating from certain Asian countries (including Indonesia, Sri Lanka and the Philippines) results in inhumane treatment towards them while they are employed in Middle Eastern states. Some of the most important issues pertaining to migrant workers are most prominent in Saudi Arabia, where the majority of migrant workers travel for employment. The policy of beheading, and widespread complaints of sexual exploitation in most Middle Eastern countries, especially in the Kingdom of Saudi Arabia (KSA), are significant facts which threaten potential migrants. As a result, deaths of migrant workers remain a common fact. Unfortunately, no coherent admonition is heard from the origin nations against this brutal practice. This is one of the most pressing issues for Human Rights. For instance, in mid-2011, 22 Indonesians in Saudi Arabia had already faced beheading (Asia Sentinel, 2011); the reasons however were not disclosed to their family members.

The flow of migrant workers from less developed countries (LDCs) to developed countries (DCs) is one of the key aspects of globalization. The term "migrant worker", as defined by the 1990 International Convention on the Protection of Migrant Workers, refers to "a person who is to be engaged, is engaged, or has been engaged in a remunerated activity in a State of which he or she is not a national" (Office of the High Commission for Refugees, 2004, mentioned in Kibria, 2004). Since time immemorial people have migrated. However, the factors that make people move have

DOI: 10.1057/9781137451187.0006

expanded so much and so quickly that the traditional notions of push and pull cannot be applied as a single factor anymore.

Scholars frequently claim that migration has long been an important livelihood strategy for the people who migrate. Whenever the population rose to such an extent that people could no longer secure a livelihood, they migrated elsewhere. This can be accurately applied to domestic migration. However the economic and social power of a migrant varies significantly and this in turn affects the choice of destination, and their cost versus benefit analysis. Currently two types of voluntary international migration occur from within Bangladesh. One takes place mostly to the industrialized West, and the other to Middle Eastern and Southeast Asian countries. Although there are different kinds of voluntary migration to the industrialized West, they are usually perceived as involving long-term or permanent emigrants. Migration to the Middle East and Southeast Asia is defined as short-term contract migration (Ullah, 2010a, 2011; Siddiqui, 2003).

Short-term labour migration is the most common form of population movement from Bangladesh, which usually entails exporting contract labour to Middle Eastern and Southeast Asian countries. Saudi Arabia, UAE, Kuwait, Qatar, Oman, Iraq, Libya, Bahrain, Iran, Malaysia, South Korea, Singapore, Hong Kong and Brunei are some of the major countries of destination. Saudi Arabia alone accounts for nearly one half of the total number of workers who have migrated from Bangladesh. Malaysia is the second largest employer of Bangladeshi workers. However, since the financial crisis of 1997, the number of Bangladeshis migrating to Malaysia has dropped drastically (see Table 2.1) and the UAE has taken its place. The systematic recording of information on the migration of Bangladeshi workers began in the mid-1970s. The Bureau of Manpower, Employment and Training (BMET) of the Labor Ministry, maintains the record. According to a Bangladesh Bureau of Manpower report in 2012 about 607,798 Bangladeshi workers have migrated to different countries of the world (BMET, 2012). The following sections illustrate the destination countries of Bangladeshi migrants.

Migrant workers in Malaysia

The social reality of migration has a long history in Malaysia; about five to six generations of migrants have impacted the structural development

DOI: 10.1057/9781137451187.0006

TABLE 2.1   *Country-wise overseas employment in 2011 (major countries)*

Country	Number of migrants
UAE	282,739
Oman	135,265
Singapore	48,667
Lebanon	19,169
KSA	15,039
Qatar	13,111
Italy	7,624
Mauritius	5,353
Brunei	5,150
S. Korea	2,021
Bahrain	13,996
Malaysia	742
Libya	89
UK	30
Kuwait	29
Others	19,038

Source: Compiled from Bangladesh Bureau of Manpower, Employment and Training (2012).

of the country over the years. Among the migrants are the early seafarers and agriculturalists, traders and merchants, evangelists, miners, explorers, colonialists, and labourers, including modern day refugees from Somalia, and stateless persons such as Kachins and Rohingyas from Myanmar. It is, however, noteworthy that these migrants have demonstrated varied reasons for moving to Malaysia, and their activities have shaped the country to a large extent. The complex and dynamic composition of migrants in Malaysia signifies that spatial relocations emerged from real or imagined, positive or negative social changes, and environmental concerns at the sending, transit and receiving areas (Abdulazeez, Bab and Pathmanathan, 2011).

Malaysia plays both sending and receiving roles in international migration. International migration is not a new phenomenon in Malaysia's immigration history. The first stage, during the 10th century AD, occurred when Malaysia was an international migration country as a result of its trade and commerce. During the second stage, in the 15th century, Malaysia became a destination for social and economic international migrants from China and India. The British colonial period saw economic conditions improve during the 1920s and 1930s, and introduced a more paternalistic policy towards the Malays. Post Second World

DOI: 10.1057/9781137451187.0006

War immigration has virtually ceased, the new independent Malaysia initiated national development programmes of import-substitution and export-oriented industrialization. The efforts commenced the process of reconstructing its national, racial, cultural, religious and economic borders and introduced further restrictions on immigration with the exception of skilled migrants.

Malaysia's migration record indicates that in mid-1990, Malaysia had 1,014,156 (5.6 percent of 18.103 million population) migrant workers, while figures rose to 1,192,734 (5.8 percent of 20.594 million population) in 1995, and again to 1,553,777 (6.7 percent of 23.274 million population) in 2000. In 2005 and 2010, the proportion of migrants in the country was estimated to have increased to 2,029,208 (7.9 percent of 25.633 million population) and 2,357,603 (8.4 percent of 27.914 million population), respectively. Although these are projected figures they indicate a gradual rise in the country's migration population. Beyond the estimated numbers of international migrants given by the UN DESA 2009, Malaysia has been battling cases of undocumented migrants (Kassim, 2005; Omar, 2005; Peters, 2005) across her states of Sabah and Sarawak, including other parts of the country that share borderlands and seas with neighbouring countries.

During the period of 1986–1997 Malaysia's rapid economic growth stimulated contemporary labour flows from ASEAN and East Asia in an unprecedented proportion. Accurate statistics are difficult to come by due to the significant incidence of illegal border crossings, as well as undocumented of exits of regularized workers. From the approximately 500,000 foreign workers in 1984, numbers swelled to in excess of 1.2 million by 1991 (Ullah, 2010a; Kassim, 2005). By the mid-1990s, foreign workers made up 1 percent of the labour force. Even after the tumultuous Asian Currency Crisis, which led to retrenchment, the official assessment was that the number of foreign workers in Malaysia had doubled to 2.4 million (Ministry of Human Resources, 2010). Foreign Direct Investment from the developed countries, and the cheap human capital from LDC countries, have been the twin pillars of high growth in the Malaysian economy. Perhaps the most important obstacle in studying Bangladeshi migrants employed in Malaysia is the insufficiency and inconsistency of the available data.

The statistics above show that in 2011 there were about 1.62 million legal workers in Malaysia. There were also more than 800,000 illegal workers living in Malaysia. The Ministry of Human Resources announced that

DOI: 10.1057/9781137451187.0006

TABLE 2.2   *Sector-wise migrant workers in Malaysia*

Sector	Indonesia	Bangladesh	Nepal	Myanmar	India
Domestic Workers	89,391	64	63	114	497
Construction	151,333	50,303	3,050	12,221	3,488
Manufacturing	127,127	144,332	209,446	116,478	7,982
Services	28,587	22,002	26,502	19,368	38,648
Plantations	214,594	20,480	2,032	3,462	16,954
Agriculture	81,777	82,294	10,323	8,861	27,543
Total	792,809	319,475	251,416	160,504	95,112

Source: Adapted from Hector (2011).

the migrant population would increase to five million by the year 2011 to keep up with the development of the country (Ministry of Human Resources, 2010). However, no data were found to that effect but it is widely believed that the number of irregular migrants was three times higher than the regular ones.

Migration to the Gulf States

In the Cooperation Council for the Arab States of the Gulf (GCC) – Bahrain, Kuwait, Oman, Qatar, Saudi Arabia, United Arab Emirates – the boom of labour migration began with the discovery of oil in the Gulf States in the 1970s. This triggered migration from South and Southeast Asia to meet the high demand of labour for construction and infrastructural development (Rahman, 2011, p. 395; Jarallah, 2009, p. 6). The rising wealth also raised the demand for household servants from Sri Lanka, Indonesia, the Philippines and Bangladesh prompting female migration into the region. Employing one or several Southeast Asian maids has become a symbol of prestige for middle- and upper-class families in the Middle East (Castles and Miller, 1998, p. 11; Malecki and Ewers, 2007, p. 476; Silvey, 2006, pp. 24, 28). Interestingly enough, the word "Filipinas" has become synonymous with "maid". Hence, it is common in the Middle Eastern countries to hear spoken in gatherings "our Filipina is from Indonesia" meaning that our maid is from Indonesia.

Middle Eastern countries remain the major market for Sri Lankan labour migrants. The Sri Lankan foreign employment industry has been growing steadily since the 1980s, and from 1995 the number of departures increased compared to the preceding years. This can be explained

DOI: 10.1057/9781137451187.0006

by the fact that in 1995 the Sri Lankan Government established a counter for registration at the airport so that workers could register immediately prior to departure (SLBFE, 2009, p. 4).

In 2009, 247,119 Sri Lankans worked abroad (119,276 male and 127,843 female) (SLBFE, 2009). Female participation in labour migration is higher than males in Sri Lanka. Since 1988, the proportion of female workers has always been over 50 percent; except for 2008 when it was only 48.81 percent. Furthermore, most Sri Lankans go abroad to work as housemaids, and Gulf countries like Saudi Arabia and Kuwait are their main destinations.

In Sri Lanka, the number of licensed agencies grew from 139 in 1985 to 746 in 2009. Almost 60 percent of these agencies are concentrated in Colombo, and another 12 percent in Kurunegala (SLBFE, 2009, pp. 83–84). Sub agents often negotiate between the agencies who are located in the cities and their potential clients living in the countryside (Afsar, 2009, pp. 18 19).

From 2006 to 2007 Bangladesh's official number of workers going abroad more than doubled. While in 2006, 381,516 Bangladeshis went abroad, the numbers exceeded 830,000 in 2007 and 2008 before falling to 465,351 in 2009 and 383,150 in 2010 after the global recession. The boost in 2007 can be explained by the construction boom in the United Arab Emirates and the lifting of the freeze on recruitment of Bangladeshi workers in Malaysia as a result of the Memorandum of Understanding (MOU), which was signed between Bangladesh and Malaysia in 2006 (OKUP, 2009, p. 22). The decline after 2008 can probably be explained by the global financial crisis, the effects of which led to decreased demand.

Most Bangladeshi labourers are concentrated in the GCC States, and Singapore and Malaysia in the Southeast Asian region. In 2010, 203,308 Bangladeshis (51 percent of total migration) were employed in the United Arab Emirates; 42,641 (10 percent) in Oman; 21,824 (6 percent) in Bahrain; 12,085 (3 percent) in Qatar; and 7,069 (2 percent) in Saudi Arabia. Other significant countries of employment in 2010 were Singapore with 39,053 (10 percent) and Libya with 12,132 (3 percent) of Bangladeshi workers (BMET, 2011).

Several studies about Bangladeshi female labour migration have been conducted. The low numbers of female emigration before 2003 can be explained by the fact that between 1997 and 2003 the government prohibited unskilled and semi-skilled labour migration of women (Rahman, 2011, p. 399). However, an estimated 45,000 Bangladeshi women have

DOI: 10.1057/9781137451187.0006

emigrated to the GCC clandestinely in recent years (Rahman, 2011, p. 399). Bangladesh in a small scale has begun to export female workers, in particular to Hong Kong and Qatar, since 2012.

Bangladeshi migrants generally take three main channels to get to their destinations: government ministries, private agencies and migrant chain networks. Today, migration is managed both by private and public agencies. However, in the early 1970s, there were only a handful of private agencies in Asia. This means that the government was the main player in labour migration. As the volume of labourers seeking foreign employment keeps growing, private agencies have flourished. However, some institutions and agencies were established to manage the migration flows in many countries, particularly in South Asia. For example, in Bangladesh five key ministries are involved in the process: Ministry of Expatriates' Welfare and Overseas Employment; the Ministry of Home Affairs; the Ministry of Foreign Affairs; the Ministry of Finance; and the Ministry of Civil Aviation and Tourism (Siddiqui, 2005, p. 4). However, complaints are widespread about the lack of coordination among the government agencies. Recently the state has taken on initiatives to export manpower to Malaysia at a lower cost than the private agencies. Private agencies are attempting to counter the government's decision.

Until 2001, the Ministry of Labor and Employment of Bangladesh was in charge of international migration for work. In 2001, the Government established a new ministry, the Ministry of Expatriates' Welfare and Overseas Employment, to respond to this emerging issue. Two main areas of activity have been established in order to promote employment creation, and to improve the condition and welfare of Bangladeshi expatriates. The new ministry is responsible for implementing the rules established in 2002 under the Emigration Ordinance 1982, including the promotion, monitoring and regulation of the migration sector. This is why the BMET is the executing organ of the Ministry of Expatriates' Welfare and Overseas Employment in respect to processing labour migration, as private agencies have to register with the BMET. Foreign missions also have important roles in regards to migration. Their tasks include exploring the potential job market, attestation of documents pertaining to recruitment, providing consular services for Bangladeshi workers, and ensuring the welfare of migrant workers.

In 1984, the government established the Bangladesh Overseas Employment Services Limited (BOESL) to take a direct recruitment role. By February 1999, BOESL had recruited a total of 8,900 workers, that is,

DOI: 10.1057/9781137451187.0006

0.31 percent of the total number of those who went overseas through the official channel (Siddiqui, 2009).

Bangladesh coped quickly with the exponential growth of global migration. As of 2009, we found there to be nearly1000 private agencies with government registration. We assume there may be at least three times more agencies working without government permission (BMET, 2009). Most licensed agencies are organized under the Bangladesh Association of International Recruiting Agencies (BAIRA), which was founded in 1984. In 2002, the association counted around 700 member agencies (Siddiqui, 2009, p. 5).

A number of studies have confirmed that 55–60 percent of Bangladeshis find jobs overseas through private contacts (Ullah, 2008, 2010a; Siddiqui, 2005). Usually, persons already employed in the host countries arrange visas for their friends and relatives through their own contacts. Those who obtain a visa through this process pay less than those who pass through formal recruiting agents. The risk of fraudulent practices is considered to be less when working through personal contracts, as opposed to the high rates of abuses within recruiting agencies (Siddiqui, 2005, p. 6). This point is particularly crucial when considering that the most dangerous aspect of labour migration occurs during the actual travel to a destination country, when a migrant may fall victim to poor conditions, trafficking, and other hazards of unregulated travel. These risks might be mitigated by the original migration arrangement set up by government agencies, individually, or by private networks (for more see Ullah, 2010a).

Many workers are unable to earn an adequate income while abroad, and are unable to repay the money they borrowed, or help their families back home (Charles, 2009). Frequently workers find themselves employed in a job that does not match their original contract. The sister of a deceased Bangladeshi worker (from heart attack) reported "My brother did not have a job for many months initially. Later, he worked for a company other than the one that originally hired him" (2009). This person also reported that her brother borrowed most of the money he needed to travel to Dubai (where he was to work) from a lender with an interest rate of 96 percent, which is to be paid in monthly instalments, leaving little room for sending money to family in Bangladesh. The Human Rights Watch Report (2006) indicated that suicides among Bangladeshi construction workers were increasing, stating that they were mostly stress related. Many of the workers were not paid properly,

DOI: 10.1057/9781137451187.0006

or sometimes not at all, and felt trapped as they were unable to pay back their loans: this led to the contemplation of suicide, as it seemed to be the only way out of what looked to them to be an irreparable situation (Human Rights Watch, 2006).

It is common for employers to hold the passports of their workers, threatening lower wages if they refuse to "surrender" them; however this makes workers vulnerable because they risk deportation if detained by authorities without proper documentation (Marriot, 2008). Whatever the norm for those employers, taking passports is illegal, and this is really the point at which exploitation starts for many migrant workers. Workers who are also forced to sign a new contract in a foreign language once they have arrived, often for a much lower wage than agreed upon prior to coming, have little ability to refuse (2008). Their immigration status is utterly dependent on their employment status, so arguing and making demands on their employer jeopardizes their employment. Once workers are held in this vulnerable position they often find themselves facing abuse from their employers. In 2002, a Bahraini man was jailed for three months after being accused of abusing and torturing a Bangladeshi employee; the employee said that the employer had tied him up with rope, beat him and gave him electric shocks. The employer's defence said that he did this in order to frighten the worker and prevent him from running away. Additionally, in 2005 the Migrant Workers Protection Society (MWPS), based in Bahrain, reported more than 20 cases of rape involving foreign house workers; however no one was ever convicted of these crimes (Ahmed, 2007). The above anecdotes demonstrate that migrants are frequently not respected in the countries they work in, nor are they adequately protected by the destination or origin country's laws.

The procedure to obtain a work visa can be very complex and it differs from case to case, for example Bangladeshi workers use three differ-ent visas to enter Saudi Arabia: (1) fixed contracts, (2) free visas, and (3) *Umra*. In fixed contracts a worker is employed in a specified job, and he or she usually enters the country on a visa tied to a fixed contract. Among Afsar's interviewees, 44 percent of the 45 male migrants and all 15 female workers went abroad with this type of visa. The second option is a free visa, in which a relative or friend acts as a sponsor for the worker to obtain a visa. The visa is not connected to a specific job or field. The third kind of visa, *Umra*, is the process by which some enter the country legally for the purpose of a pilgrimage to Mecca or Medina and then stay on for work (Afsar, 2009).

DOI: 10.1057/9781137451187.0006

The cost of overseas migration emerged as the major inhibiting factor in the migration decision-making process. The average cost of migration is estimated at Taka 144,584 for male migrants; female workers spent 2.3 times the cost for their migration. Only one woman spent more than Taka 100,000 to procure her visa. When looking at types of visas, those migrants who obtained *Umra* visas themselves paid the lowest price, followed by those who obtained a fixed job visa. The so-called "free" visa was the most expensive. The type of visa was not the only factor that determined cost. When looking at recruitment channels the lowest cost was for those who used formal channels (the Bangladesh Overseas Employment Services Limited, BOESL), followed by those who used the services of sub-agents, and then those who arranged for migration themselves. The highest costs were borne by those who relied on friends for their migration (although there were only two of these in the sample, reducing the reliability of the figures). The Government fixed the recruitment cost at Taka 84,000. However, this figure is mainly applicable for labour recruitment to Malaysia. The total cost of migration to the Gulf countries should normally be lower than this government figure, as can be seen in the case of migrants recruited by BOESL. The remainder of the respondents, however, spent more than the rate stipulated for Malaysia (Ullah, 2010a, 2013, 2014). The above sections elaborate on several reasons how the costs of migration can become a crushing burden. Loans with high interest rates, exorbitant visa costs, and inadequate protection for labour migrants all act to push migrants into a trap from which they cannot escape. This kind of situation exerts such a magnitude of stress upon migrants that it can lead to death.

Nanban quoted Malaysian Human Rights (Suhakam) commissioner Datuk N. Siva Subramaniam as saying about 1,300 illegal immigrants have died in detention during the past 6 years in Malaysia. He said many of them died in immigration detention centres, prisons, and police lockups because they were denied timely medical treatment. A total of 481 Bangladeshi expatriates died in Malaysia in 2009 alone, the Expatriates' Welfare and Overseas Employment Minister of Bangladesh told Parliament. The Minister said Taka 5.8 million was given from the Wage Earners' Welfare Fund to families of 230 dead Bangladeshi workers for transportation and funeral cost (Bdnews24, 2014). He said another Taka 8.2 million was given to the families of 51 people by way of financial assistance. (In Bangladesh, the Wage Earners' Welfare Fund for migrant workers is based on subscriptions from migrant workers, licences of

DOI: 10.1057/9781137451187.0006

recruiting agencies, surcharges on fees collected through the missions abroad, and personal and institutional contributions.) The minister also reported that the Malaysian authorities gave US$134,018 as compensation to the families of 18 dead Bangladeshi workers (bdnews24.com/sum/skb/std/1900h). In 2013, the UAE court ordered a lorry driver, who had been fined Dh 52,000 and sentenced to a year in prison for killing 21 migrants, of which 19 were Bangladeshis, to pay a further Dh4.2 million in blood money. The ill-fated migrants were travelling in a bus.

According to BMET and Zia International Airport, from 2003–2012, there were a total of 15,752 (legal and illegal) Bangladeshi migrant worker casualties repatriated. The last 3 years the number rose to 8,932. That calculates to an astounding figure of nearly eight to ten coffins repatriated daily, a figure that portrays the high cost of working abroad for too many Bangladeshi migrants working in foreign countries. The KSA has the highest number of migrant deaths (30 percent), followed by Malaysia, and then Kuwait. Road accidents (24 percent) and heart attack (21 percent) were among the most common causes of death.

Migrant worker's welfare fund

In order to support migrants who get into difficulties overseas, a number of labour-sending countries have recently established welfare funds formed through levies from migrants. The Philippines pioneered this initiative.

Assistance in forced repatriation for reasons such as illness, violation of contract, medical care, court litigation is provided from the welfare fund. This includes family members at home (Rosario, 2008). The Wage Earners' Welfare Fund for Migrant Workers in Bangladesh is also augmented by contributions from migrant workers, fees for licences for recruiting agencies etc.

In 1990, on the basis of the Emigration Ordinance of 1982, the Government of Bangladesh created a fund for ensuring the welfare of wage earners. The Wage Earners' Welfare Fund was later supplemented by one of the three rules framed in December 2002 under the 1982 Ordinance. Funds for the Wage Earners' Welfare Fund are raised through the compulsory contribution of Taka 300 from each migrant worker and from a portion of charges levied by Bangladeshi Embassies for services (attestation of documents, verification of visa, etc.) and donations (UNDP, 2002).

DOI: 10.1057/9781137451187.0006

The Wage Earners' Welfare Fund was created through a government notification dated 15 November 1990. The notification recognized that migrants are extremely vulnerable, and that there are a number of incidents that can lead to a state of utter helplessness; among those mentioned was employment termination of contract of Bangladeshi migrant workers at the place of employment, breach of employment contract, illness and absence of proper documents. In such situations, migrant workers may not have the means to return home. These and other related considerations led the government to create this Wage Earners' Welfare Fund. The stated objectives of the fund include:

(a) Establishment of a hostel-cum-briefing centre;
(b) Briefing for migrant workers by the BMET;
(c) Welfare desk at the airport;
(d) In cases of need, payment for bringing back the remains of migrant workers from foreign countries;
(e) In cases of need, payment for bringing back sick and disabled migrant workers from foreign countries;
(f) Helping families of deceased migrant workers;
(g) Providing legal aid to migrant workers through Bangladesh Embassies or High Commissions; and
(h) Establishing recreation and information centres at Bangladesh Embassies or High Commissions.

The fund is managed by a committee, initially consisting of nine members with the Secretary of the Ministry of Expatriate Welfare and Overseas Employment as the chairman and eight other members, comprising the Director General of BMET, Joint Secretary of Ministry of Labor and Employment, Director General of Ministry of Foreign Affairs, Joint Secretary of Ministry of Civil Aviation and Tourism, Joint Secretary of Ministry of Home Affairs, Joint Secretary of Ministry of Finance, Executive Director of Bangladesh Bank and finally the Director (Welfare) of BMET. Through a new notification dated the 17 February 1999, the Managing Director of Bangladesh Overseas Employment and Services Limited (BOESL) and two members of the Bangladesh Association of International Recruiting Agencies (BAIRA) were also included on the committee. The fund is yet to have any representation of the migrant workers. As of 2013, more than Taka 600 million has been collected for the fund.

In this section we have presented an overview of the migration flow, investigated labour migrants and the gaps in protection, the crisis of migrant fatalities, case study examples, international human rights, remittance, and the welfare funds. The evidence indicates that an overwhelming majority of Bangladesh's labour migrants migrate to Middle East countries. From the analysis, it is also clear that migrants are extremely vulnerable in labour-receiving countries, and even more so in the Middle East. The post-mortem repatriation of migrant labourers presents a grim picture of international migration, and it is apparent that their deaths have become a common fact. Unfortunately, remittance dependent states refuse to speak out against this crisis, and brutal practices of the death penalty. Thus this emerges as one of today's most pressing human right's issues. Although the remittances sent by the migrants contributes to the livelihood of their family and national economy, the nation fails to adequately protect them.

Legal instruments in selected countries

A coordinated and fresh legal response to the protection of migrant workers has become an important present day demand. In many parts of the developing world, in particular, the existing labour law and social protection regimes seem to have failed to offer effective responses. Hence Oliver and Govindjee (2013) propose the development of a synergized legal regime in order to address the multifaceted problems associated with labour migration. They argue that the superimposing of immigration law on the social security legal and labour law framework excludes irregularly employed migrant workers from their labour rights.

National law is also concerned with satisfying security considerations and with allowing states to retain control over the process of the immigration of foreigners. Simultaneously, however, countries are becoming increasingly cognizant of the promotion of economic growth through the employment of needed foreign labour, the facilitation of foreign investment and the increase of skilled human resources (Riedel, 2007). Immigration legislation endeavours are meant to prevent illegal immigration on the one hand, with due recognition of human rights and a human rights-based culture of enforcement, and compliance with international human rights on the other. From a human rights perspective, states must respect, protect, promote and fulfil the human rights of non-citizens, and governments that exercise their ability to defend the

DOI: 10.1057/9781137451187.0006

sovereignty of their state are required to do so in full respect of their human rights obligations to migrants.

> Non-citizens constitute an example of a vulnerable group of people, frequently existing at the margins of mainstream society and battling to make ends meet. This is as a result of evidence which demonstrates that they experience unfair discrimination and great difficulty in exercising their basic rights. Unlike citizens, migrants are generally only able to enter and live in another country legally through the express consent of that country's authorities. In essence, their vulnerability stems from past experience and knowledge of the discrimination and inequality (in terms of treatment and work opportunities) that migrants have experienced in their daily lives. (Oliver and Govindjee, 2013, p. 6)

Malaysia has a clear policy on foreign labour, and a very comprehensive range of policy instruments to control the intake of foreign labour and stem irregular migration. In practice, poor governance and inter-agency rivalry have compromised rigorous and consistent policy implementation. Exploitation and human rights violations are reportedly perpetrated by police, locals, employers, and members of The People's volunteer Corps in Malaysia (RELA) alike, which number three million strong (Ullah, 2013). Migrants face a vast range of discriminatory policies, one of the consequences of which is that employers may act unilaterally to terminate a worker's permit. Although Malaysia made an amnesty announcement in 2011 for two million irregular migrants, critics have characterized it as a move to repair their damaged reputation as a result of brutality towards irregular migrants (Allard, 2011). The stringent controls are such that some migrant workers opt to remain with an employer despite experiencing exploitation, as they may be keen to avoid legal problems. Those who choose to leave exploitative workplaces often remain undocumented and fall prey to traffickers and recruiters for an indefinite period of time (Ullah, 2013). In 1997, the introduction of thorough annual health checks for workers, and the establishment of the Foreign Workers Medical Examination and Monitoring Agency (FOMEMA), did not alleviate the troubles of migrant workers in Malaysia (Kassim, 2005).

In Saudi Arabia, the country's criminal justice system is of utmost importance when highlighting the human rights violations of migrant workers. The execution of migrant workers, on the basis of "unfair" trials, is a significant problem. The reasons for their executions are sometimes

DOI: 10.1057/9781137451187.0006

unknown by the workers, moreover their embassies are oblivious of the fact that they were sentenced to death. It is also problematic that migrant labourers are frequently not aware of their rights under the law, and are often not legally represented in the courtroom. Some have claimed to be threatened with torture if they did not sign confession statements, which they often cannot understand because the documents are drafted in Arabic rather than in their native language (Human Rights Watch, 2004).

Migrant workers lack legal protection, and their well-being and interests are frequently overlooked in the destination countries' labour infrastructure. Nowhere is this more evident than the treatment of sex workers under Saudi law which, Nabeel Rajab (2010) explains, subjects these workers to tough physical punishments. The police often do not inquire if workers were forced into prostitution or if they were acting with consent. More often than not these workers may have been raped by their employers, and then further victimized by the legal system after they were exposed to imprisonment or even sentenced to death. The kingdom's strict laws fail to take into account the particular vulnerabilities and experiences of migrant workers. Rajab poignantly states that "in reality law can be used against migrant workers, but it cannot be used by them" (Rajab, 2010).

Another case of abuse follows the story of Monir. He paid Taka 120,000 to go to Saudi Arabia; ironically he had to pay in order to escape it as well. He, as with many other migrants, ended up doing a very different job than the one that had been agreed upon before his travel. He worked as a domestic helper with no salary, no freedom of movement and no food. He was beaten, threatened with weapons, and thrown off the second floor. Monir underlined the cost of this experience, affirming that "now because of this we are under even greater financial debt" (Kibria, 2004, p. 18).

The extent of the failure for some Middle Eastern countries to extend legal protection to migrant workers is glaringly apparent from their discriminatory laws. Stefanie Grant (2005) shows how migrant workers in Kuwait and Saudi Arabia are not protected by national social and labour laws (Grant, 2005). A report presented by the International Federation for Human Rights and the Egyptian Organization for Human Rights tells us that foreign workers are not allowed to join labour unions (FIDH and Egyptian Organization for Human Rights, 2003). Discrimination on the basis of economic status is common and an issue of great concern.

DOI: 10.1057/9781137451187.0006

When the migrant comes from a low socio-economic status there is an increased likelihood of being sentenced to death (Rajab, 2010).

It is obvious that Saudi Arabian laws and policies do not meet international standards, and are often in direct conflict with international law. Stefanie Grant explains that it is illegal under international laws to subject migrants to unfair trials, and to sentence them to death without prior notification to their consular officials. Consular officials must be notified in the shortest time possible of the arrested citizen's request to contact them (Grant, 2005). The delayed, or absence of, timely notification to the migrant's consul is also illegal under Saudi law, which presents a significant contradiction. Human Rights Watch (2008) explains that the criminal procedure code in Saudi Arabia gives an arrested or detained person the right to contact anyone in order to share the information of his arrest, and the right to legal representation. Saudi Arabia also violates its international obligations, considering that the Vienna Convention on Consular Affairs includes the right of consular officials to communicate with their nationals, and vice versa (Human Rights Watch, 2008). The "rights-based approach" to migration put forward by the ILO, UNICEF and the UN Special Reporters must be properly implemented in Saudi Arabia (Grant, 2005).

One of the major challenges plaguing consulates is the inadequacy of trained staff. Grant (2005) adds that not only are consulates struggling with the lack of trained staff but also that they frequently do not have the financial resources to deal with the large numbers of cases presented to them. Grant suggests that in order to develop diplomatic protection that is founded on human rights the consulate representation must be based on the human right principle of non-discrimination. She highlights the fact that both the state of nationality and state of residence consider non-discrimination to be legally binding (Grant, 2005). The rights of migrant labourers, as Grant tells us, are included in numerous laws and treaties. Because the rights are defined and protected by numerous different legal bodies it is difficult to ensure that migrants know their rights. Likewise, protection obligations are similarly unknown to many policymakers.

The *kafala* system

The *kafala* (sponsorship) system is yet another method by which Saudi laws and policies undermine the security and well-being of migrant domestic workers by granting the employer enormous control over the

DOI: 10.1057/9781137451187.0006

worker's legal status. Not only does this system restrict movement with the receiving country, it does not allow the workers to repatriate to their home countries because they require their employers' consent to transfer from their places of employment or obtain an "exit visa" (Human Rights Watch, 2008).

The *kafala* system is the only way available to those wanting to enter and work in a Gulf country legally. This system states four different types of visas: house visa, company visa, sponsorship visa by state institutions, and sponsorship for business partnership visa. Among them the most concerning are the house and company visas, because they provide a platform for the illegal trafficking-in of persons. To enter into the country with those types of visas can be dangerous for migrant workers, and in the case of house visas (mostly for domestic jobs), workers are totally under the control of their employers, without any rights or freedom of movement. With this type of visa, sponsors can cede their workers (as if they were their own goods) to other employers and sponsors, without taking the workers' will into any consideration. The company visa, also known as the "free visa", does not offer protection to workers either. In fact, the employer is not responsible for any criminal or illegal activities that the worker commits, therefore illegal trafficking and any sort of violations can be committed by migrant workers at the sponsor's request, or from pressure or threat (IOM, 2005).

With these type of visas the sponsors have total control of their employees, there are even cases in which workers had to return to their home countries at the behest of their sponsors, without the workers having the opportunity to seek another job. One respondent, Muntasir, worked as a jewellery maker in Saudi Arabia for 3 years. He fell ill because there were no workplace safety procedures in place, and his job required him to work with acids and other chemical products. He spent all his money on medicines and medical treatments; when Muntasir complained, and asked for an increase in his salary, his employer reacted by sending him to the airport with a ticket. Muntasir was eventually beaten, and forced to return to Bangladesh. The same happened to Habib in Malaysia, he injured his hand while working; he had to pay for all the medical treatments and, even though he was able to continue his work, the employer sent him back home (Kibria, 2004).

The sponsorship system has been subjected to a review, the outcome of which was the creation of a few large recruitment agencies that would themselves be the sponsors for all migrant workers in Saudi Arabia

DOI: 10.1057/9781137451187.0006

(Human Rights Watch, 2008). A significant problem pertaining to employment contracts under the *kafala* system is that they hardly ever include specific information on work conditions, essential information like limits to the number of working hours and a proper, accurate description of work responsibilities (Human Rights Watch, 2008).

One specific grim issue that arises when migrants die in Saudi Arabia is that it is sometimes extremely difficult to send the dead bodies back to their native countries. An embassy official emphasized the constraints of the *kafala* sponsorship system in that regard, noting that the sponsor must cooperate by changing and authenticating documents in order to repatriate the body. Even accounting for these difficulties approximately 20 bodies are sent home from Saudi Arabia every month (Human Rights Watch, 2008).

Kingdom of Saudi Arabia

Many semi- and low-skilled labourers see the oil-rich Middle Eastern country of Saudi Arabia as one of the most advantageous destinations (Ullah, 2010a; Kapiszewski, 2006). Since the Saudi economic growth of the late 1960s, dependency on the foreign workforce has grown at an accelerated pace. South Asian expatriates work in low and semi-skilled category jobs such as janitors, construction workers, gardeners, salesmen, sweepers and drivers. Although there are claims of reduced foreign labour dependency, foreign workers still dominate the private sector.

Over the last decade the incidence of low-paid Bangladeshi worker deaths has been prevalently observed (Maxwell Stamp, 2010). The majority of the deaths have been found to be caused by work-related accidents. Employers are primarily responsible for bearing the cost of all compensations such as pending salaries, benefits, gratuity and other expenses associated with the worker's death. A study was carried out whereby Saudi employers and workers were interviewed so as to ascertain the actual amount of benefits, salaries and gratuities that were paid to the family members, as opposed to what was required under Saudi labour law and reported in formal job contracts.

One such incident involved a Saudi national who shot two Indian expatriate brothers following an argument at a workshop in Taif region. The brothers, Mohammed Zakhir Ahmad, 41, and Ahmad Yasin, 46, both died at the scene of the attack, a third brother escaped unhurt. The

DOI: 10.1057/9781137451187.0006

Saudi national was running a motel in the same town where the killing took place. On the day of the assault, the Saudi confronted them for not finishing the work as had been agreed upon. The brothers were shot after the Saudi took a gun from his car.

In another case international human rights organizations reported that Saudi Arabia executed eight Bangladeshi men in Riyadh. According to the available information, the migrant workers were beheaded in public for the alleged murder of an Egyptian man in April 2007. Executions in the conservative Gulf kingdom resume after the holy month of Ramadan, and Amnesty International says they are at an exceedingly high rate. "Court proceedings in Saudi Arabia fall far short of international standards for fair trial and news of these recent multiple executions is deeply disturbing" (Amnesty International, 2011).

The Saudi authorities appear to have increased the number of executions in recent months, a move that puts the country at odds with the worldwide trend against the death penalty. "The government must establish an immediate moratorium on executions in the Kingdom and commute all death sentences, with a view to abolishing the death penalty completely". According to statistics, the latest executions bring the total number in the country to at least 58, more than double last year. At least 20 of those executed have been foreign nationals (Ullah, 2011). The Bangladeshi men who were executed were Ma'mun Abdul Mannan, Faruq Jamal, Sumon Miah, Mohammed Sumon, Shafiq al-Islam, Mas'ud Shamsul Haque, Abu al-Hussain Ahmed and Mutir al-Rahman. The Egyptian man was killed during a clash between the Bangladeshi workers and a group of men who allegedly were stealing electric cable from a building complex where the Bangladeshis worked. Three other Bangladeshis were sentenced to prison terms and flogging.

The case of Kuwait

Every year the Arab Gulf States recruit thousands of workers from developing countries through their governments and recruiting agencies. Migrants are lured with offers of good wages, however when they arrive it is often a different story. Kuwait is a very small country with a population of about 2.7 million, and as it controls one-tenth of the global oil reserve, it is also completely dependent on migrant workers. On 24 April 2005, over 1000 Bangladeshi workers from a cleaning company in Kuwait

DOI: 10.1057/9781137451187.0006

demonstrated outside of the Embassy of Bangladesh demanding that the diplomats intervene on their behalf; according to police, approximately 5,000 workers had not been paid their wages for six months (Socialist World, 2005). Labour unrest has been increasing in Kuwait since 2003 due in part to the escalating level of exploitation and abuse.

Many cases of female migrants "escaping from abuses" have been recorded by other migrants working in Kuwait. One young woman called Ruama, who went to Kuwait in the late 1990s, shared her story with us. Ruama's brother arranged with an agency to work in a factory. Once she arrived in Kuwait, she discovered she was hired to work as a domestic servant in a large family. She worked under strict conditions, in charge of taking care of the children, cooking and cleaning. The hard work was coupled with abuse and poor treatment. Ruama stated that not only was she beaten and imprisoned, she risked being sexually abused by her employer, and was denied her salary (Kibria, 2004).

In 2006, Kuwait stopped hiring workers from Bangladesh, claiming that Bangladeshis were engaged in crimes. This meant that a negative image developed in the minds of Kuwaiti employers. However, many of the workers who were already there still remained in the country. In July 2008, 3 years after the cleaning company protest, 5,000 Bangladeshi workers went on indefinite strike in Kuwait city, protesting against the Kuwait Company for underpayment and irregular wages (Palma, 2008). The Kuwait Company places migrants in positions at royal palaces, hospitals, universities, oil companies and other establishments under the ministries of education, health, oil and defence (2008). According to accounts in the *Daily Star*, the company only paid the workers 18 Kuwaiti dinars a month, instead of the 50 dinars they were promised, and many of the workers alleged that they had not been paid any wages at all in two months. Some of the workers were even forced to work 16 hours per day (Marriot, 2008). Unsurprisingly many of the workers fell ill and were unable to take sick leave: a worker was quoted as saying "Most of the workers are falling sick because of the long hours of work. The company is also not allowing us to take sick leave. How can we work under such an environment?" (Palma, 2008) The Kuwait Company refused any leave to many of the workers, even to some employed for the past 8–10 years, and if they were granted leave the company charged them 30 dinars as "security" to ensure that they returned. The returning migrants rarely saw the deposits again.

Kuwait responded to this unrest by arresting and deporting over 800 Bangladeshi workers during the first few weeks of the 2008 protests. The

DOI: 10.1057/9781137451187.0006

arrested workers were ill-treated by the police; one is quoted as saying "the army beat us mercilessly while breaking up the protest, and also in (the) detention camps", the evidence of these beatings was apparent on the persons returned from Kuwait after the deportation (Marriot, 2008).

After the vocal and open response of the usually invisible migrant population, Kuwait was forced to acknowledge its dependency on migrant labour for the menial, yet important jobs within the country (2008). Supposedly, Kuwait placed restrictions on five companies accused of abusing foreign workers, preventing them from receiving government contracts. After these protests, the government of Kuwait set a minimum wage of 40 dinars, or $150 per month for the workers contracted by private companies to clean its offices, schools, and hospitals (Arab News, 2008).

As stated before, overseas workers send billions of dollars a year to their families back home; these remittances are an important addition to the Bangladeshi economy. Bangladesh's Foreign Adviser, Iftekhar Ahmed Chowdhury, wrote letters to the Kuwait Interior Minister regarding the labour unrest. Chowdhury's primary concern was that "for the faults of a few, many are being mercilessly deported empty-handed" (Bangladesh News, 2008). The Bangladeshi government was forced to realize they needed to take a greater role in protecting workers abroad, as remittances are a crucial part of the economy. If workers will not go abroad then this capital influx would dwindle significantly. The cessation of emigration would increase domestic unemployment as well. These economic gains of Bangladeshi migrant labour are self-evident, however it is also important that the government's concerns are not just driven by the risk of migrants returning "empty-handed", without money to help the economy, but rather by the rights of migrants themselves. There is no question that the number of Bangladeshi migrant fatalities would decrease with the implementation of a comprehensive programme that promoted the awareness of workers' rights and provided real legal and political protection. This goal will never be realized if Bangladeshi officials do not hold the receiving countries accountable for the abuses against their labour force.

Death penalty: a human rights issue?

The debate about the death penalty is controversial, but even those in favour of capital punishment believe that it should be practised with respect for human dignity.

DOI: 10.1057/9781137451187.0006

The death penalty in Saudi Arabia follows a precise procedure. The convict is brought into the courtyard after weekly Friday prayers with their hands tied, and forced to bow. An executioner swings a huge sword amid shouts from onlookers. There are widespread arguments in favour of Qisas, the Arabic word for Islamic-law punishments, which in the kingdom could mean beheadings or the amputation of limbs. However, many Islamic scholars counter-argue that Qisas is obsolete in the 21st century. Islam permits the death penalty for certain crimes, but few mainstream Muslim scholars and observers believe beheadings are sanctioned by Islamic law.

Capital punishment became widespread during the middle ages. From Rome to the modern era, the death penalty was applied throughout Western Europe for more than two thousand years. This topic has been debated and discussed for years. As a result of the dialogues, 96 countries have abolished it; nine retain it for crimes committed in exceptional circumstances (war crimes); 34 permit its use for ordinary crimes, but have not used it for at least 10 years and are believed to have a policy or established practice of not carrying out executions, or it is under a moratorium. Of the 50 independent states in Europe that are UN members, one country maintains the death penalty in both law and practice; 46 have abolished it; and three retain it for crimes committed in exceptional circumstances (such as war crimes). There are many conflicting arguments between pro and anti-death penalty blocks. Anti-death penalty advocates assert that they do not want to see innocent inmates executed, and have moral and religion reasons for opposing the death penalty. It has been reported that one execution in every seven involved an innocent, wrongly condemned, and most probably because of lawyer inexperience or lack of skill. Moreover, the death penalty cannot be considered as a deterrent, in fact there is no correlation between the number of crimes and the number of legislative death penalties carried out, and it obviously fails to rehabilitate. The death penalty also has a high economic cost.

Pro-death penalty advocates want capital crimes to stop, and support the death penalty with the assumption that it deters crime and saves innocent lives. Those concerned with the thought of the possibility of executing an innocent man/woman think that the death penalty should only be used on serial killers and repeat offenders.

All of these concerns are pertinent to states upholding capital punishment; however, countries that still uphold the death penalty

DOI: 10.1057/9781137451187.0006

are pondering how executions could be made less traumatic and less dehumanizing to the convicts. Lethal injection is being used in many countries, but even this method is being reconsidered due to its cause of excruciating pain. In the US it has been reported that some executions took up to half an hour; this is contrary to the purpose for which lethal injections were introduced in the first place, because it was designed to prevent many of the disturbing images associated with other forms of execution. Public stoning is another extremely disturbing practice of execution. Some believe if a person is convicted of adultery, s/he should be stoned to death. This gruesome sentence is still carried out in some parts of the world.

Thousands of dead bodies of unfortunate migrant workers arrive at airports in Bangladesh and Sri Lanka from Middle Eastern countries every year. The convicts (of 7 October 2011 in the KSA) saw no government to government diplomatic initiatives, or negotiations with the family in Egypt that could have forgiven them on blood money – sort of compensation paid by an offender or his/her family group to the family or kin group of the victim. Had it happened, these unfortunate convicts would have at least had the comfort of knowing that though their country was poor, it was a strong nation that cared for its citizens. On that day, 8 Bangladeshis were beheaded in public in Riyadh after they were found guilty of killing an Egyptian in 2007.

Malaysian case studies

This section presents a few selected case studies. These case studies enable us to understand the abject conditions plaguing migrant workers. Families of the deceased have not accepted the official cause of death (cardiac arrest) that was recorded on death certificates. Out of sensitivity and ethical concern, the narratives' participants are identified as A, B etc.

Participant A (35) came from a rural area in the Sirajgonj district, and migrated to Malaysia 15 years ago. He grew up in a typical low-income family with six siblings. The oldest sister is married, and the younger one is a student of class 9 in a local madrasa (religious school). Two of his other brothers work in different factories in Malaysia; the other lives in Bangladesh, and is a student of class 12 at a local college. Participant A lived as a legal worker in Jinjiang, Kepong, and Selangor Malaysia. He worked in a clothes shop as a salesman. For over a decade he had been

DOI: 10.1057/9781137451187.0006

working as a local salesman in different shops when he arranged his brother's move to Malaysia through legal means. Our researchers were told by the neighbours that he had been an excellent person. Everyone around there loved him very much for he never smoked, and he didn't have any other bad habits.

Participant (A) died after an attack in 2012. That night he went to a tea stall to drink tea at around 10pm. The road was dark, very narrow and isolated. On his way back he was attacked by hijackers. He tried to defend himself, but one of the hijackers hit him with an iron rod. Some neighbours observed the fight but did not intervene. He was rushed to a hospital where he struggled for life for eight days before succumbing to death. It was suspected that the hijackers were from Sabah Sarawak; however local police wanted to defer attention from the case and claimed that Participant A was pursuing a local girl, and that was why he was attacked. Long-time migrants knew that this was false, but the police refused to charge locals with a crime against a migrant, and filed a case as "unnatural death", and no one was arrested.

Within seven days of his passing away his remains were repatriated with funds supplied by friends and family. At the time of the interview, Participant A had an account containing RM1000, an amount his brothers were trying to withdraw; it is unknown whether they were successful in recovering the money from the bank.

This case may not be directly linked with the migrant's work or employer, however two important facts emerge, one is that, according to many, this death was likely a result of xenophobia, and the other illustrates the extent of a migrant worker's vulnerability that they might be subject to such a violent attack, and that no legal support is visibly accorded to him (even after death).

Participant B (30), married with one daughter, came from the Kachua thana, Chadpur district and migrated to Malaysia in 2005. He was one of ten siblings. One of his brothers also works in Malaysia, the rest are unemployed in Bangladesh. Before moving to Malaysia he used to run a petty business in Bangladesh. He mortgaged his property off, and borrowed money for migration. Participant B was working in Damansara, Selangor when he passed away. The official cause of death was recorded as a heart attack. He had been a hardworking and healthy person. One day after his return from work he fell sick. His neighbour called an ambulance that was slow to arrive. The doctor declared him dead as soon as he arrived at hospital. It is important to note here that

DOI: 10.1057/9781137451187.0006

the reason behind his sudden illness was not investigated. At the time of this study, it was not known where he had left money, nor was it known if it was ever recovered.

Participant C (25), from the Gopalgonj district, moved to Kuwait 5 years ago. There are five siblings in his family, but he was the only earning member of the family. He completed his graduation from college and tried to obtain a job in Bangladesh. When he failed to get a job, he began to plan to move out of the country. His father and brother helped him to raise the money to finance his migration. Participant C found work in the restaurant of a factory. His contract ended 3 years ago but he did not return home. He did not return because in his 3 years of service he was unable to recoup the cost of his initial migration. Though he intended to go back to Bangladesh and marry, he became an illegal worker.

In 2011, Participant C died. He was injured while working in a factory, and refused to disclose the nature or extent of his injuries. His condition continued to deteriorate. Friends wanted to bring him to the hospital, but he declined because he was scared that he may be arrested once the hospital authorities discovered his illegal status. He eventually died in a hospital where his body was kept for four days before the IOM was able to arrange for his transportation. Some of his friends claimed that he had some money deposited with another foreigner. After his death that person disappeared. This means it is highly unlikely that any of the family members of the deceased person would be able to trace and recover this money.

Participant D (40) moved to Malaysia from the Shariotpur district 12 years ago. He was married to an Indonesian woman in Malaysia, and at the time of his death she was pregnant. Participant D had seven siblings, of which he was the youngest. Eldest brother is a teacher in a local madrasa, and other brother is a farmer. Participant D migrated legally as a factory worker, and he died when a factory accident caused a blast in a cylinder. Though he was rushed to the University Malaya Hospital, he eventually succumbed to death 14 days later. Friends collected the required money to repatriate him back to Bangladesh. At the time of his passing, Participant D had a fixed deposit in a CIMB bank amounting to RM 10,000. After his death his wife gave birth to a baby girl, and was able to obtain access to the account, therefore there was no point in sending the money back home to Bangladesh.

Participant E (36), from the Munshigonj district, migrated to Saudi Arabia as a factory worker 8 years ago. Friends reported that he was confined inside the factory for days at a time, and that it was unknown whether his employer supplied him with adequate food. At the time, friends of the deceased were afraid to report the incident as they were unsure of potential consequences. Eventually he was released, but though he had been treated badly the cause of the abuse was unknown. This kind of incident, according to many of them, occurred a number of times over the course of 2 years, the result of which was that he became very ill. He was never offered proper medical treatment.

When the participant passed away in April 2010, friends were called in to the factory to move him to the hospital for an autopsy. The autopsy was performed, and a death certificate was issued stating that cardiac arrest was the cause for death. Friends claimed that at the time of his death he had an outstanding salary for five months, but that they did not know how to approach the authorities to receive the money. Interviewers spoke with the family to ascertain if they were able to recover the wages; the family claimed total ignorance about it. Friends are not entitled to retrieve the money. Families generally are not given information of the woes they go through abroad.

All of the case studies share a common theme: the fact that no one in any of the five cases received aid from the High Commission of Bangladesh. They just received a death certificate to release the body. Officials of the High Commission revealed that before 2009 they used to cover the costs for migrants who were indigent. The Commission use to repatriate the body, or send money for treatment; this stopped after 2009. Though the welfare fund still exists, the High Commission does not use it to the benefit of migrant workers or their families. When any Bangladeshi migrant worker dies while working in Malaysia, their friends collect the funds for companies to repatriate their remains. In most cases, there are claims that Bangladesh's High Commission employs non-career diplomats who are appointed on the basis of political affiliation rather than merit. They clearly do not serve the cause of the migrant population.

A way around the deaths

Not all migrants are fated to suffer the same abuses and neglect as the sad cases outlined above. Let us offer some of the concrete policies established by specific companies in Brunei and Saudi Arabia that have

DOI: 10.1057/9781137451187.0006

been successful. Al Raed Architecture and Consultants to date have not experienced any fatalities in their workforce. It is however, explicitly mentioned in the job contract that in the case of a worker's death the employer shall be responsible to send the deceased to his/her country of origin along with the pending salary, gratuity and other benefits to the deceased's family. It is important to note that there is no third party involved in such monetary transactions. All necessary and official arrangements are made through the company, so that the pending salaries along with other monetary compensations are wired directly to one of the family members' bank accounts. It is for this purpose that at the time of signing the contract, Al Raed collects detailed records of the employees' emergency contacts. As provided by the employer, the contracts between the employees and employers clearly portray the responsibilities and duties of the employer in the case of the death of any recruited employee. An excerpted section of the contract between the employee and employer is shown below:

> "In the event of death of the employee during the terms of this agreement, his remains and personal effects shall be repatriated to the country of origin at the expense of the employer. In case the repatriation of remains is not possible, the same may be disposed of upon prior approval of the employee's next of kin/or by the employee's embassy/consulate nearest the jobsite."

Most of the employees working in this consulting company are architects, draftsmen, civil engineers, electrical engineers, public relation managers, business analysts and other office supporting staff. Due to the nature of their desk job, exposure to the death risk caused by extensive physical labour is significantly less.

The Rashed Al Rashed and Sons Group also has a policy in place regarding the death of employees. They actually implement Saudi's Labor Law as it pertains to migrant deaths and family benefits. The process starts by verifying the causes of death, whether it is caused by a work-related accident or it is a natural death, that determination allocates the amount of monetary benefits that are to be paid to the deceased's family. The verification process includes a number of stakeholders, such as the police department, and the court (Sharia law), into their determination.

The company's research participant was interviewed, and asked a set of questions in chronological order. The first question was "Irrespective of the cause of death, generally what is the amount of money and benefits being handed over to the employees in the status of construction workers?" The

DOI: 10.1057/9781137451187.0006

manager replied that the company does not have any independent policy concerning this but that every compensation and benefit paid is determined by standards set in the "Labor Law" of Saudi Arabia.

For both of the following questions the human resources manager referred to the Saudi Labor Law, that is, pertinent actions are taken in accordance with the laws depicted in state labour law:

> Question: In the case of work exposed death, where the deceased worked as a construction worker, generally what are the benefits (in monetary value) being handed over to the deceased's family?

In response to the following question: *"What is the number of death events of the employees from, for example, Philippines, Bangladesh, Pakistan, India, etc.?"* The respondent mentioned that the Rashed Al Rashed Company has not yet had an employee death, but he stated that if it occurs company policy is to follow the law of the land.

When asked: *"What are the major steps involved in verification and decision on the benefits for the deceased?"* the manager explained that the entire process entails a number of necessary steps. First, the death has to be reported to the local police, and the concerned embassy of the deceased. Upon reporting, the local police department takes charge to investigate and verify the cause of death. Afterwards the police department sends the death report to the court for further action and decision. Once the court has received the report, it starts the process to determine the amount of money that must be paid to the deceased's family.

The average duration from the point of verification to the payment of benefits is approximately one to six months. The duration of the process is affected by the cause and circumstances of each case. In the absence of emergency contact information, the company interviews the closest colleague or friend of the deceased to get detailed contact information regarding their family.

The Al Saud Consult is one of the largest consulting firms in the Eastern province, offering consulting services for planning, design and construction projects. It is typically engaged in large construction projects throughout Saudi Arabia. Like the companies mentioned above, Al Saud Consult follows a strict procedure in the event of an incident of death, however also like the above companies, they have yet to experience a death of an employee. In the event that one should occur they state that the "Labor Law" of Saudi Arabia is the primary and only legislative deed used to determine the amount of money to be paid to

DOI: 10.1057/9781137451187.0006

the deceased's family in the country of origin. The interviewee was asked same set of questions, and the responses were similar to the Rashed Al Rashed and Sons group.

Emaar Middle East is a construction outsourcing company that finds specialized contracting firms to fulfil construction orders. Emaar directly engages workers for construction, however, they do ensure that the contracting companies perform their jobs and maintain standard working conditions for the lower-level employees (e.g., construction labourers) in compliance with the Labor Law of Saudi Arabia. According to Ahmad Alkadi, no death case has been reported thus far. .

Unlike other companies, Saeed R. Al-Zahrani Corporation (SRACO), has reported a number of incidents. Three Bangladeshi janitors employed by SRACO were interviewed. All of them provided detailed information on different death incidents that occurred in different years. Table 2.3 below shows a summary of the workplace-related SRACO employees' deaths.

It can be seen that SRACO managed to send all the pending salaries and gratuities owed to the deceased's families via the coffins of the deceased. However, it was not evidenced that the deaths were verified by the legal standard procedure in Saudi Arabia. Also, it was not possible to know whether SRACO has any record of following the Saudi Labor Law while determining the amount of benefits eligible for the deceased's family. Their practice of sending the money inside the coffins brings into question the level of professionalism and fulfilment of responsibilities by SRACO.

Foreign workers in SRACO are predominantly employed as janitors, gardeners and sweepers. As a part of the job contract SRACO provides a very ordinary accommodation facility in a labour camp located in Dammam, called "SRACO Labor Camp", which is also popularly known as the "Camp". In the camp one driver is specifically employed for emergency and regular transport of workers to workplaces, hospitals, or the airport, etc. When a deceased worker (*Mohammad Jainal*) was feeling excruciating stomach ache, he informed that camp driver to take him to the hospital. However, the driver showed his utmost reluctance and totally ignored the request, imposing an excuse that the pain was caused due to ordinary causes. The driver asked him to go to bed for a rest. The older man was not persuasive, because of the pain, and went back to his room quietly. When getting to the room, he asked his roommates for a glass of water, which he drank before becoming very calm and quiet. His

DOI: 10.1057/9781137451187.0006

TABLE 2.3 *Profile of three migrants who died in KSA*

Employee's Name	A	B	C
Working Station	Dhahran	Dhahran	Dhahran
Age	36	25	50
Home District	Manikganj	Brahmanbaria	Brahmanbaria
Causes of Death/Time	Health hazard e.g., gastric, ulcer, etc.	Road accidents	Health hazard e.g., gastric, ulcer, etc.
Date of Death	August 2010	May 2011	February 2010
Family Members	Wife and 2 children	5 brothers in the family	Wife and 4 daughters
Pending Money Status	Pending salary was handed over to the deceased family along with the coffin carrying the dead body.	Pending salary was handed over to the deceased family along with the coffin carrying the dead body.	Pending salary was handed over to the deceased family along with the coffin carrying the dead body.
Money Borrowed or Lent from Friends/Colleagues	His colleagues forgave him for the money (usually small amount) deceased borrowed.	His colleagues forgave him for the small amount of money deceased borrowed.	His friends and colleagues forgave him for the money deceased borrowed.
Relation with Interviewee	Co-worker of the deceased	Co-worker of the deceased	Co-worker of the deceased

Source: Field study (2012–2013).

DOI: 10.1057/9781137451187.0006

roommates then realized that Jainal's condition had already deteriorated, and they became furious with the camp driver who eventually drove the ill man to the hospital. While on his way to the hospital Jainal died. Due to such gross negligence, and other proven misdoings, the camp driver was fired and deported to his country of origin.

The accounts of the janitor respondents are clearly varied from the types of interview responses gained from the other professional firms. Their status as low-level workers employed in a company like SRACO adds authenticity to their account; however their understanding of the incidents were limited by their position. For instance, it was impossible to ascertain the exact figure of benefits and back pay supplied to the deceaseds' families because the janitors simply did not have access to that information.

During our research we managed to get a SRACO worker's official contract paper with the Bangladeshi counterpart company called "Musa International". It is interesting (and this has been a common practice when it comes to Bangladeshi workers employed in low category jobs) that the actual reported salary for the SRACO workers is 300–350 SAR/month, which is significantly less than the 550 SAR/month which was originally agreed upon. This could be grounds for scepticism regarding the actual amount of money given to the families of the deceased by SRACO because they failed to follow the procedural outlines of labour protection laws. Unfortunately, none of the casualties above were verified by the local police department, or the monetary benefits determined by the court.

DOI: 10.1057/9781137451187.0006

3
Profiling the Deceased Migrants

Abstract: *Chapter 3 presents an analysis of the profiles of the migrants who died in the destination countries and their families living in Bangladesh in order to understand the context in which Bangladeshi worker fatalities occur. The chapter discusses the morbidity pattern and the demographic and socio-economic histories of the concerned families with a view to unearthing the level of vulnerability that they are exposed to after the death of a loved one. Putting forward the horrifying statistics of death casualties in the destination countries, the authors have evidenced the unrealistic claims as to the causes of death of the migrants in linking them to parental history. This chapter also touches upon an array of important facts, for example how much money was left at the destination, what mechanisms were used to recover the money, and how the families were compensated by the government or employers, if at all.*

Ullah, AKM Ahsan, Mallik Akram Hossain and Kazi Maruful Islam. *Migration and Worker Fatalities Abroad.* Basingstoke: Palgrave Macmillan, 2015. DOI: 10.1057/9781137451187.0007.

In order to understand the context in which Bangladeshi worker fatalities occur, it is necessary to analyse the demographic and socio-economic characteristics of the migrants who died abroad. This chapter analyses these socio-demographic and economic features of deceased migrants, including age, family size, number of children, marital status, the main sources of household income, land ownership and major professions of deceased workers. This enables us to better understand the dependency of the families on migrant workers, and how much more vulnerable they become after the only income earner has passed away.

The field data collected between 2010 and 2013 show that the majority of the migrants who died were in an age group of 20–40, and married with at least one or two children. Generally, age, family size and number of siblings are related to why people leave their countries. This of course reminds us of the Ravenstein model (1880s) that postulates that migration is mostly undertaken by people at a young age. Family size and number of siblings are explained by household strategy theories, meaning that in order to diversify the income source of the family, people tend to migrate.

The age distributions of the deceased workers are presented in Table 3.1 below. Among the total of dead migrant workers identified (76) around 4 percent died very young (less than 20 years of age), the broader concentration of deaths is on the second range where 44.7 per cent died at the age of between 20 and 30 years, followed by 38.2 per cent between 30 and 40 years. These figures show the cumulative percentage of 86.8 migrants who died at 40 years of age or younger. The mean age of

TABLE 3.1   *Distribution of Bangladeshi workers died abroad (by age)*

Age of workers (year)	f	%
< 20	3	3.9
20–30	34	44.8
30–40	29	38.2
40–50	7	9.2
50–60	2	2.6
> 60	1	1.3
Total	76	100.0

Source: Field data (2010–2013).

DOI: 10.1057/9781137451187.0007

the workers was 33.5, with only 25 per cent still unmarried. The table in fact means that we were able to identify 76 families that lost their members while abroad.

The majority of the deceased migrants had a fairly smaller family size as compared to the national standard. This has crucial implications in understanding the dependency ratio (DR) in the family. The DR is important in order to comprehend how dependent the entire family had been on the migrant member. An overwhelming percentage of the migrants (52.6 per cent) came from joint families. The workers' family members most commonly number between three and five members, which means that most probably the workers are either married with at least one child, or if they are not married, they have parents and other relatives to take care of. About 21 per cent of the families consist of three members, 17 per cent of four members and about 18 per cent are formed of five members. Fifty per cent of their families range in size from two to four and the rest ranges from five to 12.

Fifty seven (75 per cent) of them were married and had children, and only 5 (6.6 per cent) did not have any immediate dependents, meaning that the overwhelming majority had wives and children to mourn their death. It was very apparent that the death of their family members pushed them entirely into a life of uncertainty. Their financial support had been jeopardized, family bonds shattered, and the future education of their

TABLE 3.2 *Family size*

Family size	f	%
2	9	11.8
3	16	21.1
4	13	17.1
5	14	18.4
6	6	7.9
7	6	7.9
8	5	6.6
9	3	3.9
11	3	3.9
12	1	1.3
Total	76	100.0

Source: Field data (2010–2013).

DOI: 10.1057/9781137451187.0007

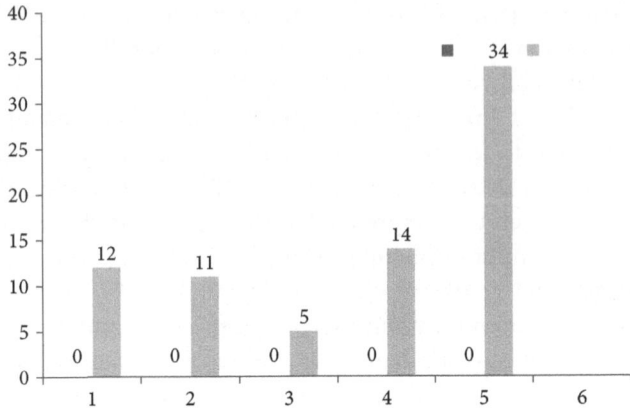

FIGURE 3.1   *Number of children*
Source: Field data (2010–2013).

school-aged children had fallen apart. There were many bereaved families that were kept waiting to receive the money the migrants may have left in the destination countries. One mother of a dead migrant lamented:

> ...my tears dried up, strength got exhausted, my heart became dismantled, but still I was not told what was my son's fault, why he had to die an untimely die,...who will take care of us? Who will call me to enquire about my health...

Socio-economic characteristics

The death of the migrant not only stopped the income to the family, but it also deprived them of the security he or she used to provide. This placed the families into both financial and social insecurity. The study revealed that about 25 per cent of the families were left with no earnings. Of the other 75 per cent, 51 per cent had someone economically active but no income. The source of the families' income was dominated by agriculture (26.32 per cent), then labour (11.84 per cent), remittance (6.57 per cent) and other mixed sources (27.63 per cent). Other sources of income reported were tailoring (3.96 per cent), business (5.26 per cent), and savings in banks or elsewhere (1.32 per cent). The figures establish the fact that most of the workers' families relied on the agricultural sector, which is a typical feature of rural Bangladesh. About 54 per cent of the families got all or part of their income from agriculture. This is of

DOI: 10.1057/9781137451187.0007

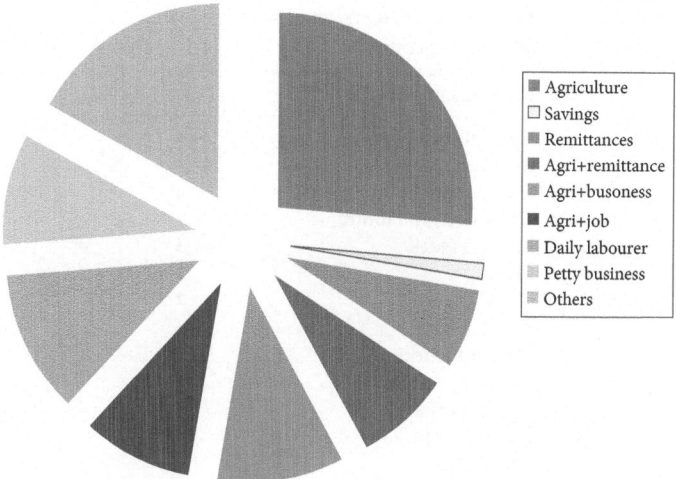

Agriculture
Savings
Remittances
Agri+remittance
Agri+busoness
Agri+job
Daily labourer
Petty business
Others

FIGURE 3.2   *Current livelihood strategy of family*
Source: Field data (2010–2012).

particular concern because the majority sold out their significant assets to finance their family members' migration. In most cases it is impossible to recoup those losses, therefore the labourers' death put the families into a multi-tiered predicament. Death of an earning member of a family pushes down their social status. Many family members reported that neighbours became paranoid about the death of their family members. Some outspoken neighbours declared that he was either a criminal or a drug addict, and that was how death occurred. These kinds of statements are not only pejorative toward a deceased but also devastating to those who are alive.

Agriculture is the main economic sector on which Bangladeshis and their economy rely; it is imperative to further analyse the relationship between the first sector and land property. The situation seems more dramatic when we take into consideration the fact that nearly half of the families involved in agriculture do not own any land. This means that their subsistence is volatile, and subject to change. Those who reported to have a piece of land also reported that it was too small for them to rely upon for subsistence.

Another relevant element to analyse is the previous profession of workers. Almost 85 per cent of migrants were involved in some kind

DOI: 10.1057/9781137451187.0007

TABLE 3.3   *Profession of workers before leaving the country*

Previous professions	*f*
Agriculture	23
Business	2
Carpenter	4
Construction labour	2
Electric work	1
Fishing	1
Handloom	2
Horticulture	1
Migrant worker	1
Masonry	18
Other works	1
Poultry farm	1
Shopkeeper	2
Study	3
Tailor	1
Trolley driver	1
Unemployment	12
Total	76

Source: Field data (2010–2013).

of work while in Bangladesh. The majority of them were involved in agriculture-related activities while 16 per cent had no job.

Bangladeshi labour migrants are mostly unskilled. Around 86 per cent of the workers did not have any training before their departure. Approximately one-third of them were offered work in masonry activities, and less than 4 per cent were employed in agriculture in foreign countries. Table 3.4 below illustrates the particulars of who was employed in which sector. This may be explained by the fact that the kind of jobs they were offered of course involved risks to health and life.

Incidence of morbidities

There is a correlation between work environment and morbidities in any context. Work environment determines the level of health risk and policy determines the management of health risk. Therefore, if policy

DOI: 10.1057/9781137451187.0007

TABLE 3.4   *Type of work offered to them in foreign country*

Type of work	f	%
Masonry	27	35.47
Labour	22	28.95
Carpenter	3	3.95
Gardening	3	3.95
Security guard	4	5.27
Electronics	1	1.32
Mechanic	2	2.64
Cleaner	2	2.64
Painter	1	1.32
Agriculture	3	3.95
Laundry	1	1.32
Driver	3	3.95
Total	76	100.0
Others	4	5.27

Source: Field data (2010–2013).

as well as work environment are poor, migrant populations face greater health risks than ever. The very common complaints are that that their work-related environment is unsafe, unhealthy and unhygienic in most cases. In our field visits we noticed the similar conditions in our target countries. This is not just an issue for the migrants and their advocates, it also has a negative effect on labour productivity as well. The Middle East as well as some Southeast Asian countries have gradually gained an image whereby a most unsafe and unhealthy environment is offered to migrant workers. It has been characterized by the so-called 3D condition – dangerous, dirty and demeaning. Media reports in May 2009 told of a factory in Saudi Arabia where six Bangladeshi migrants were burnt alive in their workplace (The Prothom Alo, 2009). Fire fighters arrived only after they had died. This incident illustrates that working places often do not meet the required international standards for safe working environments. There were even some instances of fires in factories in Saudi Arabia where law suits were filed against dead migrants for property damage. The claim being that the factory caught fire due to the negligence of the workers.

The survey results found that the overwhelming majority of labourers were in good health before leaving the country. About 90 per cent of respondents said that migrants did not suffer any disease before departure compared with 10.5 per cent responding that they had minor health complaints. About 60 per cent of the respondents claimed that they had telephone conversations with the migrants a week before news of their death was received. All of them said that no health problem was reported. This is the reason why family members refute the causes reported in the official death certificates. Some family members have been very critical about the statements from the government of Bangladesh. The relevant minister kept saying that those deaths were natural.

A plausible explanation for these contradictory data can be the possibility that workers do not admit their real health condition for fear of being less employable, or that they did not complain about their health problem for fear of a salary cut. These fears are not unfounded, the majority of jobs are in agriculture and construction, and they demand hard work. In either case, this is a human rights violation. Employers are quite aware of the extremely harsh conditions in which their employees live and work and most likely have reservations about hiring workers with health complications. Many migrant workers during our interview period mentioned that employers can buy their way out of legal difficulties but "we cannot".

One theme that emerged during our research was that some employers and friends abroad, questioned whether the migrants who died carried any inherited diseases from their parents. It is not unfair to suppose that this may be the case in some instances, for there are many genetic diseases that are passed on in families from one generation to the next, that is, vertical transmission. For example, whereas children may inherit physical features, such as brown eyes or curly hair from one or both parents, they may also inherit certain disorders. Some of the common hereditary diseases include heart disease, diabetes, cancer, stroke, mental illness, thyroid problems and hypertension. This study attempted to find a correlation between the recorded disease of the migrants and those of their parents.

It is evident that almost 70 per cent of the parents did not suffer from any diseases at the time of departure. More than one-third of the parents were affected by different diseases including heart disease, diabetes, asthma and other conditions, indicating that it was less likely that they died due to diseases inherited from their parents. We sought opinion from an experienced medical doctor on this issue. The doctor was surprised at our intention to correlate the death of migrant workers with their parents' diseasesand didn't provide an excuse. Most importantly, they were medically fit before they flew, he added. We attempted to delve

DOI: 10.1057/9781137451187.0007

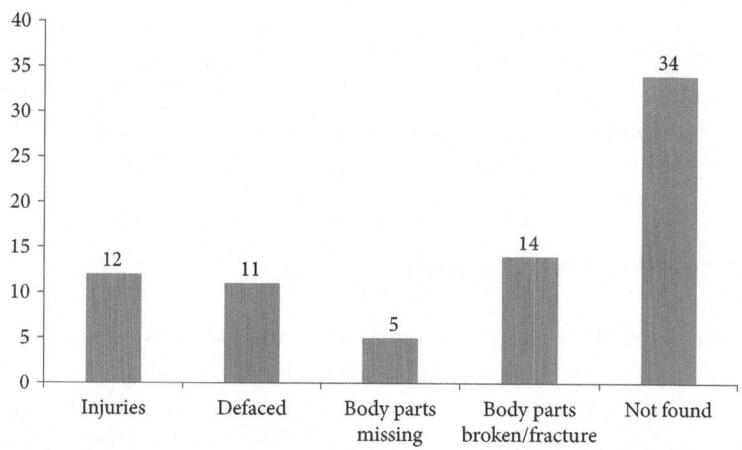

FIGURE 3.3   *Parental disease history*
Source: Field data (2010–2013).

into this fact further. We found that the rest of the siblings of the families (reportedly those whose parents had those kinds of diseases) were quite healthy.

Foreign workers in the construction sector work in some of the most difficult conditions. Working in extreme heat and a humid climate can be dangerous. In the United Arab Emirates, the government prohibited work during the hottest hours of the day in July and August in order to minimize heat-related illness (Rahman, 2009). Employers are obliged to continue to pay the salary of an injured worker during his/her treatment for up to six months. However, on anther occasion, we got to talk to at least 80 workers in the UAE and such prohibition remains on paper, but not in practice. Also, there are no comprehensive data about the rate of injuries and deaths among foreign workers.

Usually, receiving countries in the Gulf region require foreign employees to obtain health insurance. In Lebanon, for instance, health insurance is compulsory in order to obtain a work permit and visa (OKUP, 2009). In the United Arab Emirates, employers have to provide insurance for their low-skilled workers. However, many employers buy insurance only once without renewing it after one year (Rahman, 2010). The law requires the insurance of workers except when companies provide for their own medical institutions, licensed by the Abu Dhabi

DOI: 10.1057/9781137451187.0007

General Authority for Health Services, to offer medical services (Human Rights Watch, 2009). The common complaint is that migrant people do not receive adequate medical attention. Problems such as long waiting times, ignoring emergency cases, separating the labourer from their own citizens and language barriers, among others, are common.

Perceived causes of death

The increasing volume of the repatriated migrants has become a great concern for the government of Bangladesh. Newspapers report that almost every day a migrant worker's remains are returning home in a coffin. Not counting Malaysia, the field data show that the majority of the dead arrive from Middle Eastern countries, with Saudi Arabia sending the highest percentage (40). Cardiac arrest is the main reported reason for death, and is caused by three relevant issues: stress, workplace environment and poor alimentation. Most of the respondents and some other migrant workers mentioned that employers reported "cardiac arrest" as a cause of their death because this kind of fabricated report will help them win in the case of a law suit. However, the question remains why they had cardiac arrest? How could the employers not take responsibility? Cardiac arrest does not occur out of the blue. Some claims that the tension caused by income uncertainties and unhealthy food habits are also responsible for heart attack fatalities. The migrant workers in the Middle Eastern countries also eat more meat and other fatty foods than their traditional vegetable-based diets, and such food habits increase the risk of heart disease.

Job insecurity and workplace accidents are considered to be the main reasons for migrant deaths. Job insecurity is mainly correlated with poor working conditions. Migrants often work on the sweltering exterior of high-rise buildings without any sun or heat protection. Their socio-demographic status as unskilled Bangladeshi labourers makes such poor working conditions the social norm. Tension at work may also be a leading factor in the labourers' deaths. For example, as tensions rise and workers become stressed they may be unable to work attentively and a workplace accident may easily occur. A stressed and worried worker cannot focus on his/her work, and distraction at a dangerous workplace can cause a fatal accident. According to cardiologists, mental tension is a real factor in heart disease, and may contribute to the death of migrant workers. Bangladeshi labourers often suffer from acute mental tension.

DOI: 10.1057/9781137451187.0007

They work far from their relatives, isolated linguistically and culturally in their host countries. Mental tension is also caused by low income, debts, poor salary, uncertainty of jobs, and non-fulfilment of contractual obligations. It is not surprising that some are unable to cope with the unusual pressure. Exploitation is not the only stressor for workers, living abroad unaccompanied can cause homesickness – a common complaint among migrant workers. Homesickness is a mental state that can lead to inattentiveness to work. This means that they might find themselves in a complex predicament leading to permanent mental anxiety.

Some factories do not maintain a consistent break for breakfast and lunch. This leads to a habit of having untimely meals, putting them at risk of a permanent stomach ulcer and related diseases. Consistency in salary payment determines food quality. Poor eating habits are related to cardiac arrest fatalities. Not only do migrants eat more meat, and other fatty foods, but also in some cases, migrant workers do not have access to healthy food choices. Deprivation may be accentuated by actual maltreatment. We have explored abuse in the workplace, but often migrants are exposed to abuse in other domains as well. Many labourers are detained by immigration police in the Middle East and RELA members in Malaysia; there have been numerous reports of torture at the hands of security officials and forces (Ullah, 2013). The emotional and physical scars caused by torture can contribute to an early death.

Another major cause of migrant deaths is the lack of medical care in the workplace. The migrant workers we studied were medically fit at the time of leaving home, and there are no data to suggest a genetic component to their sudden illnesses. Therefore, it may be assumed that bad food habits, poor working conditions, mental stress, and inadequate medical care could all be contributing factors to the migrants' deaths. Road accidents are a leading cause of deaths worldwide, and are also responsible for many migrant deaths. Car accidents happen everywhere, even more so in countries with under-regulated traffic laws. Bangladeshi workers may engage in unsafe driving by habit, and this negligence may lead to the majority of Bangladeshi workers' road accidents.

Deaths are also caused by the Gulf Countries' capital punishments. When labourers are accused and convicted of serious crimes such as rape, hijacking, or murder they will inevitably face the death penalty in their host country. If we place in order the causes of death among workers the first cause is cardiac arrest at 43 per cent. Other diseases that can cause workers' deaths total 13 per cent. More than 56 per cent of

DOI: 10.1057/9781137451187.0007

the deaths are caused by the workers' health conditions. As we discussed before, very few workers arrived in the host country with health problems, therefore, the living conditions of the host country have a direct and negative effect on the workers' health. The second most common cause of fatality is workplace accidents, at 30 per cent of the total. Health conditions and accidents together total 86 per cent of workers' deaths.

Tables 3.5 and 3.6 show that cardiac arrest is one of the major causes (40 per cent) of deaths among the migrant cases studied. Workplace accidents totalled about 21 per cent of work related deaths, and are the second most common cause. Road accidents claimed 13 of the workers' lives, and the other reasons that we have explored account for the remaining 22.4 per cent.

The survey results show that the overwhelming majority (90 per cent) of migrants who were repatriated post-mortem were not suffering from any disease when they left their home country. The field survey revealed that about 79 per cent of the family members agreed with the reason of death specified on the official death certificate. Though there were divided opinions within the family. Some members of a family agreed while some others did not. On the other hand, more than 20 per cent (21 per cent) of

TABLE 3.5   *Number of deaths by different causes (January–May 2009)*

Causes of death	f	%
Cardiac arrests	391	43.25
Workplace accidents	268	29.65
Road accidents	62	6.86
Illness	115	12.72
Other reasons	68	7.52
Total	904	100

Source: Field data (2010–2013), and Zia International Airport Office, Bangladesh (2012).

TABLE 3.6   *Causes of death according to death certificate*

Causes of death	f
Heart attack	30
Road accident	13
Accident in workplace	16
Other	17
Total	76

Source: Field data (2010–2013).

DOI: 10.1057/9781137451187.0007

the family members did not believe the reported causes for death. There was a firm belief among those family members that their relative died because of mental and physical torture or such as sleep deprivation, mock execution and solitary confinement. The cause of death reported on the death certificates did not represent the whole truth. This implies that the foreign company where the workers were employed may tend to conceal the true cause, therefore abdicating the need to pay compensation or accept other forms of legal responsibility.

Post-mortem examinations are used to investigate a death, and their report is the pre-eminent evidence in determining the cause of a death. It is not surprising that the destination countries have an interest in recording a cause of death because the outcome of such reports deter-mines not only the legality of the labour practices but also the level of compensation owed to the family of the dead migrant. Unfortunately, during the field survey we found that the workers' families frequently complained that they encountered problems with understanding the death certificates, or holding it accountable, because they were mostly provided in the language of the destination countries. This of course resulted in the inevitable misunderstanding and miscommunication between bereaved families and employers. However, the Bangladesh government in a way facilitated the widening of the gaps in available support for migrant workers. Bangladeshi missions abroad could have played a significant role in lessening these gaps, said many.

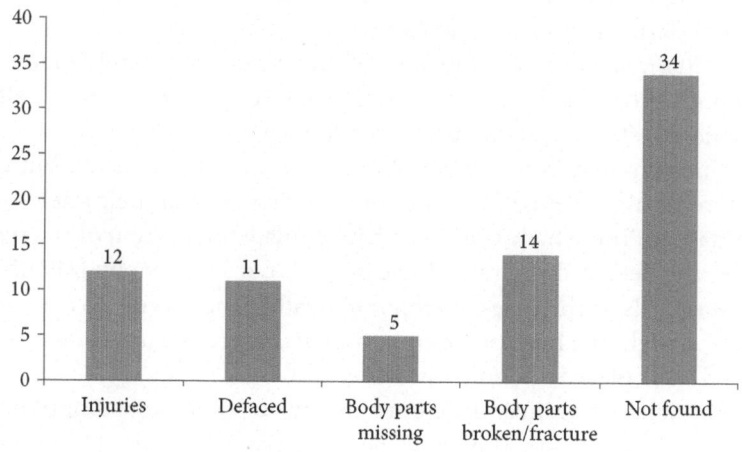

FIGURE 3.4 *Post-mortem results*
Source: Field data (2010–2012).

DOI: 10.1057/9781137451187.0007

Family members reported that they believed those death certificates to be false, concealing the reality that led to the death of their loved ones. Some relatives arranged to do a post-mortem examination once they received the dead bodies. In some cases, the post-mortem examination results conflicted with the data recorded on the official death certificates issues in the destination countries. There were stark differences between the causes reported on the certificates and the results found during post-mortem examinations done in the home country. Incidents of injuries and disfigurement of body parts were extremely high, which illustrated the sub-par and dangerous working conditions experienced in the destination countries.

The companies and authorities in the destination countries seem hesitant to acknowledge the true cause of migrant workers' deaths, therefore undermining the need to compensate the deceased's family. Bereaved families face a mountain of challenges when attempting to recover lost funds or demanding compensation; the lack of information, coupled with corruption in both countries and unscrupulous companies working for their own self-interest, deter families from taking the initiative to claim their rightful compensation (Ullah and Hossain, 2013a). The following section discusses remittance issues, including frequencies of remittances sent and the amount of the remittances. This is presented to demonstrate the pattern of their level of earning and attachment to their families in Bangladesh, indicates how much dependency the families have on the remittances and further shows that families must be in the know about their genetic health status in order to clarify their loved ones' health status.

Remittances have become one of the most important engines of economic growth for many countries. About 10 per cent of Bangladesh's GDP is constituted of migrants' remittances. It took a long time for policymakers to accord proper attention to migration issues, but once the foreign aid pipeline started to dry up, they turned their attention to remittances. There has never been any estimate the portion of the remittances the deceased migrants leave behind, nor is there any initiative to recover funds in the cases of emigrant deaths. The economic cost of the deaths is high; the human costs are horrific. Therefore, a succinct policy should be in place to handle this issue.

Remittances significantly elevate family income, consumption and savings (Ullah, 2010a). The Bangladesh Bureau of Statistics (BBS) survey shows that the incidence of poverty is 61 per cent lower in families that receive remittances. About 13.1 per cent of households receiving

DOI: 10.1057/9781137451187.0007

remittances are below the poverty line compared to the 31.5 per cent national level average (BBS, 2010). This implies that there are at least seven million families that are directly dependent on remittance.

We attempted to understand the average income that the deceased migrant normally sent home per month. Data show that around half of the respondents used to send between US$60–110, though they used to remit irregularly. Most families of the deceased believe that their loved one had left money abroad. However, because of lack of information, they were unable to recover that money. Our field study has revealed that the remittances sent by migrant workers contribute significantly to the livelihood of his or her family, and to the well-being of their country.

About 66 per cent of the workers were able to remit regularly to their families. Among the deceased workers, 71 were able to send money, and five failed partly because of the shorter period of stay or lower than expected income. More than half of the deceased migrants sent money amounting to between 5,000 and 10,000 Taka per month, and about 87 per cent sent money ranging from between 2,000 and 15,000 Taka per month. Finally, it was found that the percentage of workers who sent remittances that value 15,000 Taka per month or more was 12.6, while only 2.8 per cent sent 2,000 or less.

TABLE 3.7   *Money sent per month by the workers from abroad*

Amount sent per month (in thousand taka)*	f	%
< 2	2	2.6
2–5	10	13.2
5–10	38	50.0
10–15	12	15.8
15–20	3	3.94
20–25	1	1.3
25–50	3	3.94
50–100	2	2.62
Subtotal	71	93.4
	5	6.6
Total	**76**	**100.0**

Note: *A thousand taka is equivalent to US$13.
Source: Field data (2012).

DOI: 10.1057/9781137451187.0007

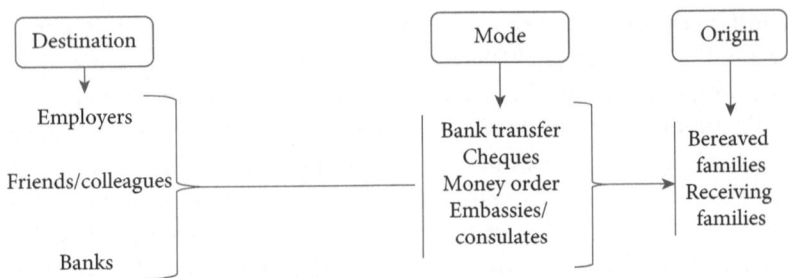

FIGURE 3.5   *Routes remittances get to receiving families*
Source: Ullah and Hossain (2013a).

The following diagram shows how the deceased migrant's assets, if any, were delivered to their bereaved family.

As described before, remittances are one of the main reasons why people migrate to another country. Remittance affects poverty most directly because it has the potential to increase a household's income. In fact, "income from remittances is usually larger than that which could have been earned by migrants had they stayed at home [therefore] migration can reduce poverty" (Buchenau, 2008, p. 4). The equation of course may not be as simple as it seems. One reason a person opts for migration is in order to obtain financial independence and encourage their personal development. However, nearly 90 per cent of Bangladeshi migrants admit to sending remittances to their families at home, although part of that money is usually used to pay back the initial travel loan. It is estimated that average Bangladeshi remittances total US$1,138 per year (sent in four times per year) (Orozco, 2010). Remittance constitutes a significant portion of the GDP of many countries in the world. A crucial developmental effect of remittances is that the longer people receive money, the lower their dependence becomes over the long term. This most likely means that families try to invest that money, and develop an alternative source of income (Orozco, 2010). This confirms all the other findings analysed so far in this work. Families that are dependent on remittance can be very vulnerable to external shock, such as the death of the one who remits, such an event can have devastating effects on their financial security.

Migrant workers' families use part of their remittance for some sort of investments, like buying land and building a house or home renovation. This is not surprising given the starting hypothesis, whereby we in

DOI: 10.1057/9781137451187.0007

TABLE 3.8   *Income and remittances by sending country*

Country of destination	Local household income	Remittances	Combined income	Per cent dependency on remittances
Kuwait	79,678	115,840	195,518	59
Saudi Arabia	57,323	109,567	166,891	66
UAE	57,933	92,554	150,488	62
Malaysia	60,390	75,323	135,713	56

Source: Compiled from IOM (2009); Orozco (2010, p. 11).

fact assumed that one of the workers' main reasons for leaving was to help their families and improve the living conditions of their relatives at home. The most common initial investments involved buying land, or buying back land they sold before in order to finance their travelling costs. Once they have bought a piece of land, the next logical step is to try to capitalize on it, so they start some income-generating activities like agriculture and/or farming. For those who are not interested in land, they prefer to buy a house or improve the construction of the one they already own. The next type of investment involved setting up a new business in order to have a more stable alternative source of income (different from the first sector). Once they have stabilized their conditions, they pay off the loans they have incurred in the migration process. Finally, with any remaining funds they try to buy a means of transportation and help other relatives and friends who desire to move abroad. Travelling abroad is expensive for migrant workers. The costs for travel, visas and work permits, housing and government fees in both the sending and host country can quickly mount up. The government of Bangladesh authorized the cost of travel to the Middle East as about Taka 40,000 to 50,000 (approximately US$540–US$680), however they often end up paying up to Taka 200,000 (about US$2,700)[1] to recruitment agencies. Many of them borrow the money, often at a high interest rate, to finance their migration (Ahmed, 2007).

Note

1   This information is from a 2007 article, but we used current exchange rates for the US$ currency value.

DOI: 10.1057/9781137451187.0007

4

The Price Migrants Pay, and Policies in Place

Abstract: *Chapter 4 attempts to charter the range of the migrants' vulnerabilities to existing migration policies at both points of the migration process. At the onset, the dynamics of abuses and deaths of the migrants and their impact on the families are explored. Focusing on the evolution of migration policy of South Asia, including Bangladesh, this chapter critically discusses the factors shaping the migration policy of Bangladesh. How the institutional framework of international migration is governed is also examined in this chapter. The authors, through this chapter, have made a strong argument by saying that although migration is structurally essential for many of the growing economies in Asia, the existing legislation of Bangladesh is inadequate to address the challenges that migrants encounter. The chapter wraps up by looking at policy challenges and reviewing the laws and initiatives of receiving countries.*

Ullah, AKM Ahsan, Mallik Akram Hossain and Kazi Maruful Islam. *Migration and Worker Fatalities Abroad.* Basingstoke: Palgrave Macmillan, 2015. DOI: 10.1057/9781137451187.0008.

DOI: 10.1057/9781137451187.0008

When performing a cost and benefit analysis of migration it is the general practice to only consider measurable monetary benefits or costs of the process. This practice is myopic, and ignores the human condition that drives migration in the first place. We have discussed at length how the gaps in protection and lack of welfare lead to social, physical, and psychological deficits, and that is a very real issue that should be of concern to all of us. Evidence abounds that low-skilled workers are subjected to sexual and physical abuse; they are deceived and not properly paid for their services. The costs of labour migration frequently mirror the costs of human trafficking; the two are closely linked. We explore the human costs of migration, and how individual migrants make the choice to move.

Every human behaviour has social consequences; when an individual makes the choice to move abroad for employment they must weigh and face these potential adverse consequences (Ullah, 2013a). Migrants are fully aware of the potential high personal costs of migration, however if they thought the costs outweighed the benefits, perhaps they would not go. Clearly, many assume that they will gain more by leaving. The moral and psychological costs incurred when leaving their home country, being far from family and friends, adapting to a new environment, embracing new cultures, and sometimes even suffering humiliations and exploitation, should not be underestimated (IOM, 2004). Even when migrants make calculated choices, knowing all the facts, their status as non-citizens in the host country prohibits them from being able to enjoy full civil and political rights while living and working abroad (Martin, 2010).

The first thing to analyse is the potential individual well-being, then the impact migration has on families and social groups and then the macro perspective of what migration means on a national level. The impact of migration is heterogeneous, thus the costs and benefits are not easy to analyse. The implications do not only extend to the labour market; social and economic policies can be hugely affected, with both positive and negative consequences. Emigration from developing countries can affect growth, education, well-being, culture and wealth distribution, Diaspora, brain drain, unskilled manpower exploitation and irregular migration (De Haas, 2009).

Migration can be a double-edged sword; it can achieve development and economic growth goals, social dynamism, diversity and cultural richness. Intercultural bridges can be built through increased workplace

DOI: 10.1057/9781137451187.0008

solidarity, multinational trade and labour agreements, and revised international business practices. All this can be established and nurtured through migration. It poses a policy challenge because the new migratory situation makes it necessary "to find the point of equilibrium between facilitation and control" (IOM, 2004, p. 9). The point of intersection where the advantages of migration meet minimal costs is where we must focus our analysis. It is irrational to imagine the world without large-scale migration. This is not just an economic issue; it rather encompasses the complex issues of human rights, public health, national identity and international security as well.

This chapter elaborates upon the hardships migrants encounter overseas. They face gender discrimination, coercive contractual circumstances, violence, and forced labour, to name a few. Migrants are also burdened with unexpected working conditions, racism, lack of health insurance, and general physical and legal vulnerability while in the host country.

Except for women working as domestic servants, the SLBFE and BMET do not indicate exactly which types of jobs migrant workers obtain. Instead, they categorize the levels of employment as "professional", "skilled", "semi-skilled", "less skilled" or "unskilled". Only a small fraction of Bangladeshis go abroad for professional work, the majority are categorized as "less skilled". There are more reliable figures on Sri Lanka's migrant workers, most of who work as housemaids. Sri Lanka also has a great number of emigrants working in the "skilled" or "semi-skilled" sectors such as construction (Palma, 2008), plantation agriculture and manufacturing. Bangladeshi women recently have started to obtain more positions outside of domestic work, as garment workers, hospital cleaners, nurses and other skilled and semi-skilled jobs (OKUP, 2009).

Migrant domestic house workers remain the most vulnerable sector of the migrant labour force. Gender plays an intricate role in this vulnerability to human rights abuses; not only are domestic workers most likely young females, the historical expectation of female domestic workers is that they somehow lose their individuality and "belong" to the house, this is especially true in Gulf States.

Saudi Arabia has an extremely poor record on women's rights. Saudi Arabia has reservations to some international treaties (Human Rights Watch, 2008). The Saudi State made it clear that it was not obliged to implement terms of the Convention on the Elimination of All Forms of

DOI: 10.1057/9781137451187.0008

Discrimination against Women (CEDAW) when they were in conflict with the norms of Islamic law (Human Rights Watch, 2008). Both the sending and receiving countries affect women's lives in different ways. Silvey (2004) asserts that a corrupt state bureaucracy is the major contributor to women's rights violations.

Human Rights Watch's investigation of abuses against domestic workers in Saudi Arabia was published in a July 2008 report, "As if I am not a human". It revealed cases of forced labour, trafficking, slavery and slavery-like conditions, alongside the more widespread complaints of non-payment of wages and long working hours. Abuses against domestic workers in Kuwait are similar to those found in Saudi Arabia, and the Kuwaiti government inadequately investigates and prosecutes allegations of abuse against domestic workers (Human Rights Watch, 2009). The case of Haima illustrates this negligence: she was a migrant worker from the Philippines who was subjected to trafficking and slave-like conditions. Her relatives had given her false information about her job abroad before she left. She was sexually harassed by her agent, and then threatened by forced return if she complained. She was subjected to sexual assault by her employer, who held her passport and forced her into workplace confinement (Human Rights Watch, 2008). Considering the different abuses and exploitation FDWs are specifically subjected to, including the withholding of wages, Halabi places special emphasis on the vital role that international pressure plays in bringing about significant policy changes to protect migrant workers (Halabi, 2008).

Women are also at risk of prostitution, either as a result of trafficking or after running away from their place of employment without their passport. Significant numbers of Indonesian maids end up as prostitutes in Kuwait, followed by Sri Lankans, Nepali and Indian migrant workers. According to Dr Trilochan Upreti, Secretary in the Prime Minister's Office at an interaction on UN convention on migrant workers' rights, "hundreds of Nepali women are trapped in prostitution net operated by crime syndicates" (NMN, 2010). This means that they become more vulnerable to abuse than before because they lose all kinds of protection nets. They remain under the control of a syndicate, and live to serve it. One of our studies shows that the likelihood of these women being victims of abuse is eight times higher than for those who are under a legitimate system (Ullah, 2003).

DOI: 10.1057/9781137451187.0008

Dynamics of abuses

The lack of government regulation and controls allows employment agencies to have undue power to determine the fate of migrant workers. Recruitment abuses, including providing false information regarding work conditions, the exorbitant recruitment fees that result in indebtedness, threatening to terminate a contract earlier than usual, and furnishing an embellished contract all ultimately amount to forced labour (Human Rights Watch, 2008).

We have seen how employers in the Gulf States have enormous control over their workers. There are widespread accounts of Saudi Arabian employers not allowing workers to contact their families; reports about physical and sexual abuse, and insufficient food supplies are common complaints against employers in Saudi Arabia (Human Rights Watch, 2008). There are exceptions of course but these exceptions are not generalizable. One of our ongoing researches confirms that at least 75 per cent of the foreign domestic helpers in the Middle East experienced mild to violent levels of abuse. Employers routinely use confinement in bedrooms and bathrooms as a form of punishment, or as a means to stop workers from escaping. Human Rights Watch reports that one domestic worker was confined in the bathroom for about eight months whenever her employer was out of the house (Human Rights Watch, 2008). There are many instances of migrant workers jumping out of windows of the employer's house in order to escape abuse. When they do escape and decide to contact the police they are often arrested. Romina Halabi (2008) tells us that escaped workers are often detained for running away or for lying to the police. Adhikary and colleagues (2011) highlights discrimination, language and cultural barriers, legal position, and low socio-economic status as important contributors to health problems of migrant workers. They explain that female migrant workers in the Middle East risk being subjected to different forms of abuse, including physical, sexual and verbal. Migrants' lives are also put at risk by accidents in the workplace as well as the spectre of suicide (Adhikary et al., 2011). In the case of LG Ariyawathi, a Sri Lankan maid who worked in Saudi Arabia, the couple she was working for hammered heated nails and needles into her body. The Sri Lankan Deputy Minister of Economic Development stated that the government was going to report her case to the government in Saudi Arabia and provided her with compensation (Jakarta Globe, 2011).

DOI: 10.1057/9781137451187.0008

It is evident that much still needs to be done in order to protect migrant workers in the Middle East. It is now time to see what the respective governments do rather than what they say. It may be safe to say that international pressure is potentially the primary and longer-term solution to the migrant workers' problems. This is especially true when we consider that bilateral agreements are rarely successful, primarily because of the power imbalance between host and sending countries. However, international pressure has limits once it clashes with the issues of economics, and is rarely enforced on a domestic or national level. Specific means of cooperation, and pressure by national and international NGOs or the global civil society, can reinforce international law.

There are other important factors that contribute to migrant protection, and they must be carefully considered. Education is arguably the most important factor as it plays a crucial role in making migrants aware of their rights; it must also aim to make sure migrants are well versed in the laws of the land in their host countries. When migrants cross borders as skilled and semi-skilled workers with higher levels of education it is inevitable that they have higher levels of protection. This has been the case with female migrant domestic workers from the Philippines, who are usually better educated than their counterparts from other Asian countries, including Indonesia, Sri Lanka, Bangladesh and Nepal. In most parts of the world the work environment for migrant workers is considered to be a non-issue; this is where the most egregious abuses occur. Some argue that abusive behaviour towards others (FDH etc.) is a reflection of their own society. The argument perhaps meant to say that wife battering and abuses are common in that society.

There is no exception in the case of Jordan. The Phoenix Centre for Economic and Informatics Studies found that around 10,000 attendants and cafeteria staff in the health support services sector were denied basic rights such as annual leave, health and safety conditions, and minimum wage, as they are guaranteed under the labour law. The study by Godfrey et al. (2004) in Kuwait showed that a major problem indicated by both male and female workers was physical and verbal abuse. Similar findings about abuse of domestic workers are provided in the case of the United Arab Emirates (Sabban, 2004), and US reports for Bahrain (US Department of State, 2010b), Jordan (US Department of State, 2010c), and Lebanon (US Department of State, 2010d).

The hardship which Bangladeshi women faced in countries of employment in the Middle East, include irregular payment of salaries, long

DOI: 10.1057/9781137451187.0008

working hours, physical and sexual abuse. Eventually, frustrations and anger develop among some workers leading to violent encounters that often result in fatalities. Among female migrant workers, pregnancy is a major concern. Pregnancy without being married is considered a crime in most Arab countries and abortion is equally prohibited. To avoid jail, pregnant domestic workers might take the risk to undergo unsafe, clandestine abortions, which may lead to death. One of the studies conducted by Ullah (2010b) on premarital pregnancy among domestic workers in Hong Kong demonstrates that an alarming number of FDHs experience premarital pregnancy and face severe attendant consequences such as repatriation, loss of job, salary cut, rebuke and shame, among others, and although Hong Kong is considered to be an open society, still most of the pregnant women reported less tolerance (Ullah, 2010b).

An Indonesian migrant worker was tortured to death in Kuwait in 2010; the public blamed the Indonesian government of being negligent, and the director of the NGO "Migrant Care" accused the Indonesian Government of not thoroughly pursuing investigations (Prameshwari, 23 July 2010). When a sending country remains silent in the face of such blatant abuse it encourages employers to continue their illegal practices.

It is a similar situation in the United Arab Emirates. There have been incidents of factory worker unrest in the UAE and Kuwait since 2006. As we discussed earlier, tens of thousands of foreign workers protested against low wages and poor treatment. Their protests were violently and brutally suppressed. Thousands of workers were deported and suspended from future recruitment (Human Rights Watch, 2009). On March 2008, around 1,500 Egyptian migrant workers working at an electromechanical plant in the Salaa area of Sharjah protested against low wages. In October 2009, "thousands of construction workers in Dubai's Jebel Ali free-trade zone smashed police cars and blocked traffic. Within weeks, about 40,000 migrants in Dubai had staged strikes to demand pay rises, including for work building Burj Dubai, the world's tallest skyscraper."[1] Foreign workers again went on strike in 2010; the majority of workers were employed in the cleaning sector, and protesting against insufficient or non-payment of wages. The Kuwaiti Government responded by barring two cleaning companies for three months from bidding on government contracts because of a wage dispute that culminated in an October cleaners' strike affecting public sector buildings (US Department of State, 2010a). These incidents speak volumes about the validity of long-standing claims by foreign workers that they have been exploited.

DOI: 10.1057/9781137451187.0008

The sponsorship system binds a worker to his or her sponsor, even prohibiting domestic workers from changing jobs without their sponsoring employers' consent (Human Rights Watch, 2010a). In Kuwait, leaving a job is illegal and means criminal penalties and deportation for "absconding" even if the worker had left his or her work to seek redress from abuse (Human Rights Watch, 2010a). They also face a fine of up to KD 600 (US$2,077), or imprisonment for up to six months prior to deportation. Such policies are obviously aimed at deterring workers from reporting workplace abuse.

Workers who leave their jobs become undocumented, without a sponsor, and are vulnerable to escalating forms of exploitation (Rahman, 2010). In Jordan, foreign workers do not have a statutory right to remove themselves from hazardous conditions without risking the loss of their jobs (US Department of State, 2010c, pp. 36–37). This makes them "more susceptible to dangerous conditions" (pp. 36–37). In Saudi Arabia, foreign workers are formally allowed to leave dangerous working conditions, but in reality they are not able to do so because employers generally confiscate their travel documents (US Department of State, 2010d, p. 48). In the United Arab Emirates, a federal law from 1980 stipulates that in the event of breach of contract by the employer, an employee is allowed to quit his or her job without prior notice. However, prior to 2005, construction workers were prohibited from transferring to other sponsors, "leaving it unclear whether they could seek work after quitting an abusive employer" (Human Rights Watch, 2009).

In the UAE workers must have spent at least 3 years working for their sponsor (according to the date their labour card was issued) before becoming eligible to request the patron's consent to transfer sponsorship, which they may do only once "during their tenure in the country" (Articles 2 (4) (c), Ministerial Decision No. 826–2005). Executive Regulations for Labour Sponsorship Transfer do not provide criteria prohibiting the sponsor from arbitrarily or unreasonably withholding consent, requiring the labourer to pay a fee for this consent or excepting workers whose patrons cannot be found. Workers who obtain such consent must pay a fee of between 5000 and 6000 dirhams (US$1362 to US$1622), which is the equivalent of roughly seven to ten months' wages for a construction worker (Article 3). For an additional fee of 3000 dirhams, labourers can transfer one year's service (Article 4). The original sponsor's consent is not required if the new one pays all the required fees, and shows

DOI: 10.1057/9781137451187.0008

that the previous one had not paid the worker for three consecutive months (Article 6). This has now been reduced to two months following a 2006 Prime Ministerial Decree that ordered the Minister of Labor to pass reforms allowing workers "who have been cheated on wages or simply not paid for more than two months" to be released from their employer sponsorships (Human Rights Watch, 2009). In Saudi Arabia, migrant labourers must reside in the area of residence of their immigration sponsor (a company or an individual) and they are restricted in finding new work or leaving the country with an exit visa unless their sponsor agrees to such a change (Human Rights Watch, 2010a, pp. 11–12). These examples all examine how policy is a root cause of exploitative labour practices; in the Gulf States it creates an environment where abuses flourish by forcing migrants to be totally dependent on their employers and not holding companies and individuals accountable for their abusive treatment.

A former Bangladeshi diplomat to the Middle East pointed out that many "workers do not know anything about the procedure of coming to the Middle East. They just sign contract papers here (Bangladesh), or in the receiving countries, without knowing anything" (Chowdhury, 2008). Low-skilled labourers must be supplied with adequate information about their host countries, its laws, and general practices as well as the general information about the migration process. Once they are employed overseas and their employers have their signed contract, it becomes notoriously difficult to escape the situation.

Initiatives undertaken to increase awareness and knowledge within the migrating population seem to have remained ineffective in many cases. In Bangladesh, the BMET does try to increase awareness and knowledge through briefing programmes for workers recruited for employment in Saudi Arabia, Malaysia, Kuwait and the Republic of Korea. With limited range of information covered, these briefing programmes last only two hours, (Rahman, 2010) which turns out to be ineffective.

In Kuwait, it is mandatory for domestic workers to sign a copy of a government-issued standardized employment contract, which is available only in English and Arabic. Migrants are often not informed of its content prior to their departure, and many workers report that they were told that their rights were vastly more expansive then those outlined in the document.

In Bahrain many migrant workers also remain unaware of their rights despite reported efforts by the Government of Bahrain to raise

DOI: 10.1057/9781137451187.0008

awareness. With all these efforts under way, why would the contract not be translated into Bengali so that the migrants could take it upon themselves to understand the terms and conditions? Is this an oversight, or an intentional ploy to keep migrant workers powerless and submissive? Some workers have reported to Human Rights Watch (2009) that they were forced to sign new contracts upon arrival.

In the UAE, the employer's confiscation of passports is encouraged by the government's policy of cancelling the visas of employees who have absconded, and their requirement that the passports be turned in to the authorities (Human Rights Watch, 2009). The recent 2009 law reforms in Bahrain and Kuwait that relax the requirement for employees needing the consent of their employer to switch jobs does not cover domestic workers (US Department of State, 2010a, p. 26).

With the exception of Bahrain (Jarallah, 2009), countries in the Middle East put domestic labour outside the jurisdiction of their labour laws (Godfrey et al., 2004; ILO, 2011b, 2011c). Recent efforts by the Jordanian Government to introduce standardized working contracts, forbidding the withholding of salaries and confiscation of passports, do not include domestic workers (Frantz, 2008). Similarly, in February 2010 a new law by the Kuwaiti Government granted extended rights to private sector employees, but it also does not cover domestic workers (MOSAL, 2010, Art. 2). In Bahrain and Lebanon, domestic workers are not allowed to join unions (US Department of State, 2010a).

OKUP (2009) noted:

> There is no mechanism in place to redress the sexual abuses and exploitation of domestic workers. The existing policy of the recruitment of domestic workers makes no mention of any remedy to protect domestic workers from such abuses and exploitation. Nor are there laws in the destination countries that protect the domestic workers from violation of their rights.

In Bahrain, foreign workers constitute approximately half of the population, and are routinely deprived "of many fundamental legal, social, and economic rights" (US Department of State, 2010a, p. 19). For many foreign workers in GCC States, it is hard to report abuses or escape dangerous and abusive working conditions. In some countries, labour migrants can seek help from their embassies. However, the legal procedures are often drawn out, and require the claimant to be willing to wait for weeks and even months in embassy shelters during the negotiation process with

DOI: 10.1057/9781137451187.0008

their sponsors (Human Rights Watch, 2010b). In Lebanon, there are no effective complaint mechanisms in place; judicial procedures are lengthy and visa policies are restrictive, discouraging many workers from lodging complaints at all (Human Rights Watch, 2010b). In Bahrain, foreign workers are allowed to become members of unions, but are not authorized to engage in collective bargaining (US Department of State, 2010a). In the United Arab Emirates, the lack of trade unions and independent NGOs "has produced a situation where the government and the business sector are the sole entities deciding on labour-related issues" (Human Rights Watch, 2009, p. 63). Workers are often afraid to unionize due to their fear of being fired. The laws of the country do not protect the right to organize, and strikes are prohibited. In Kuwait, all workers, except domestic workers, are allowed to join unions. However, as a prerequisite they have to "obtain a certificate of good conduct and moral standing from the government" (US Department of State, 2010a, p. 23). Foreign workers are not granted union voting rights, or permitted to obtain leading positions within unions. Technically they are only allowed to join unions after 5 years of residence; however, this regulation is largely not enforced. Non-domestic and maritime workers are not allowed to organize, or associate.

In Lebanon domestic workers, day labourers, temporary workers in public services, and agricultural workers in specific categories, are not protected by the country's labour laws since they would have to obtain residency before being covered. Only foreigners with legal residency can form and join unions (US Department of State, 2010a). We have examined the many inhuman sufferings of foreign migrants, and their restricted access to remedy. Their sufferings are incalculable – gender discrimination, recruitment abuses and workplace confinement, exploitative conditions, violence, vulnerability, coercive contractual circumstances, lack of healthcare, and forced labour, to name but a few. With the exception of Bahrain, countries in the Middle East keep domestic labour beyond the jurisdiction of their labour laws. The countries that do have labour friendly policies lack the initiatives for implementation. It is not an exaggeration to state that the plight of the average migrant worker constitutes a breach in human rights. Evidently, there is extremely limited space for the migrant workers to freely express their grievances, to report human rights violation and to seek redress contributing to the fatalities.

DOI: 10.1057/9781137451187.0008

Abuses and death: implications on the families

When a sole bread winner dies the living standards of their families deteriorate further. Not just because the income flow stops, but because their demise brings multiple tragedies to the families. In most cases (Ullah, 2010a, 2012, 2012a) migrants borrow money from a number of sources, sometimes with high interest rates. When they die, they leave the burden of the loan on the family. Most of the families are not in a position to repay those loans; many of them mortgaged off their properties to finance the migration in the first place with a hope that they would take the property back once remittances started to come in.

The cycle of migration, and its relative costs, can be divided into five phase. The pre-migration phase relates to the work that workers do in Bangladesh; their salaries can be between 3,000–8,000 Bangladeshi Taka (BDT). Migration is the second phase, it involves researching financial sources in order to fund the cost of migration, and can cost between 100,000–300,000 BDT. Generally, the family helps with these costs (including relatives who already live abroad); funding can also be met by selling lands or through a loan. Migrants pay monthly interests rates of up to 10 percent for loans they have taken out to fund their migration (Martin, 2010). The third stage is during the first year abroad, when migrants work in the host country and they attempt to recover the cost of their travel. As we have seen in the previous chapters, workers often do not receive the expected salary or are even employed in the expected job; therefore, the expectation to be able to resolve their debts is not met. The next stage is the "phase II abroad"; it generally starts when the initial costs have been recovered and migrants are able to save some money or invest. These savings are estimated to be around 20,000 BDT up to 25,000 BDT. The final stage of the cycle occurs when the migrant returns home to either retire or invest their new skills and assets in new income-generating activities (Buchenau, 2008).

In some cases migrants take more than 5 years just to recover their debts. In other cases they cannot afford to repay them at all, and this forces them to stay in the host country to continue working – either legally or illegally. Generally, one-third of migrant workers obtain a visa through a recruiting agency, while 68 per cent through relatives or friends who already live and work in the host country (BMET, 2008). An obvious cost of migration is related to the charges incurred in sending the

DOI: 10.1057/9781137451187.0008

TABLE 4.1   *Costs and benefits of migration*

| | Migration Phase I | | |
	Case I (Migration through family member abroad)	Case II (Using services of recruiting agency)	Phase II
Cost of migration	100,000	200,000	
Migrant's income	12,000	12,000	12,000
Expenditures family at home	4,000	4,000	4,000
Expenditures migrant abroad	(3,833)	3,238	4,000
Recovery of migration cost (w/o interest)	4,167	4,762	N/A
Savings/investment capacity	0	0	4,000
Number of months to recover migration cost	24	42	N/A

Source: Adapted from Buchenau (2008, p. 7).

remittances home. The cost of remitting money includes the direct costs for the service providers sending the money, such as fees and exchange rate differentials, and the handling or processing costs of the transaction and transportation. The transaction costs include the transportation cost to the agent abroad, while transportation costs are related to the delivery of the money to the recipient in the home country (Buchenau, 2008). The above discussion is highly relevant to deceased migrant workers, as in the aftermath of their death these circumstances put the entire family into a vicious circle of ordeals.

How much money do migrants spend on their migration? The World Bank estimates that on average, Bangladeshis spend US$2,300. This is nearly double the amount of US$1,200 that the Bangladeshi government determined. Their estimate includes: visa and other permits, the cost of obtaining a passport, airfares and other agents' fees and taxes. Obtaining the correct documentation is expensive, as is domestic transportation in order to obtain the papers. The procedure itself can take up to seven months (Ullah, 2010a; *The Financial Express*, 2009). The World Bank's experts estimated that the average cost of migration is nearly five times Bangladesh's per capita income. Of course given the high level of corruption, fraud and illegal market occupied by syndicates, the real cost is inflated and difficult to ascertain. The study found that they spent an astonishing amount of money to finance their migration. More than 90 per cent spent between the range of 50,000–300,000 Taka and

DOI: 10.1057/9781137451187.0008

TABLE 4.2   *Cost to finance migration*

Taka (00,000)	*f*	%
< 0.50 lakh*	1	1.3
0.50–1.00	12	15.8
1.00–2.00	33	43.4
2.00–3.00	25	32.9
3.00–4.00	1	1.3
4.00–5.00	2	2.6
5.00–6.00	1	1.3
6.00–7.00	1	1.3
Total	76	100.0

Note: *One lac equals to US$1,250.
Source: Field data (2010–2013).

43 per cent spent between 100,000 and 200,000 Taka and almost 33 per cent of the workers spent between 200,000 and 300,000 Taka. Despite the fact that only 6.5 per cent of workers spent more than 300,000 Taka, the average cost for the visa and travel was 200,000 Taka, which is much higher than the market price.

Visas are sold to labour migrants at an exorbitantly higher price by the manpower recruitment agencies. Most of the potential visa buyers have difficulty reading the contents in a visa. A survey conducted by the IOM (2002) states that only 50 per cent of people are aware of the regulation which sets a maximum price for agents to charge. As we have extensively discussed, lower-skilled migrants generally travel to either the Middle East (especially the Gulf countries) or Southeast Asia. Deportation is another painful reality for migrants working in the Gulf, especially during times of slow economic growth. During an economic boom, Gulf States are generally more tolerant with immigration (legally or not) whereas during times of austerity, immigrants are the first to pay the economic and social costs of the crisis.

Post-burial remittances and compensations

There are two main factors that can make the families of dead migrants financially whole. One is tracking down and obtaining the cash or assets left behind in host countries after a migrant has passed, and the other

DOI: 10.1057/9781137451187.0008

is enforcing the applicable compensations, either from the employers or from the Wage Earners' Welfare Fund in the home country. The repatriation of a migrant's remains is categorized as compensation, however we would argue that this not compensation, but an obligation of the employers. Compensation must be defined as something, typically money, awarded to someone in recognition of their loss, suffering or injury.

The survey found that 55 per cent of the families received the cost of transporting the remains home. The actual fare of the coffin is not the only cost: 12 per cent of the families reported paying bribes in order to retrieve the bodies at the airport. Only 12 per cent (9 families) of the families have been able to receive any compensation from Bangladesh's welfare fund (see Ullah and Hossain, 2013a). There are many factors that contribute to this low outcome, but the lack of information about the procedure is the most common reason families fail to receive compensation.

There are many complicated steps involved in settling the deceased migrant's affairs. The family has to trace the debts incurred in the home country, and uncover the assets that may have been left behind at the destination. In the questionnaire, the interviewees (family members of the dead migrant) were asked how they repaid the loan that was taken out during departure. Some of them reported that they had not, and were currently hiding from collectors, many more of them were trying to comply by selling whatever they had left to sell. This gives us another understanding as to why sending siblings to school is a secondary objective for migrants. This increases their families' future earnings security, which adds to the odds of survival and the ability to repay the loan. Table 4.3 represents the amount of loans taken out by a family. They are substantial. Around 10 per cent had loans below BDT 25,000, and around 23 per cent had loans between BDT 50,000 and 400,000.

Data clearly show that there is a marked decline in children's school attendance in the deceased family. This could be caused by a series of factors. The cost of attendance is high, and it is an extra expenditure that depends on the migrant's remittance. There are also the emotional dependencies that suffer when a loved one dies, the child loses a mentor and guide, and does not want to return to school. The most unfortunate cause of a decline in attendance is when the family is unable to maintain a level of sustenance without the child contributing to the families' earnings. According to the survey, sadly the children that cannot attend school

DOI: 10.1057/9781137451187.0008

TABLE 4.3   *Amount of money borrowed by the workers' family*

Amount of loan (Taka)	*f*	%
<25,000	7	9.2
25,000–50,000	8	10.5
50,000–100,000	4	5.3
100,000–150,000	1	1.3
150,000–200,000	4	5.3
200,000–300,000	1	1.3
300,000–400,000	1	1.3
Sub-total	26	34.2
No loan	50	65.8
Total	76	100.0

Source: Field data (2010–2013).

are mostly forced to work in order to bring food to the family's table. Overall, after the death of a migrant earner the entire family's consumption (on food, media, cloths and medical needs) becomes dramatically reduced. This is not an isolated event, rather it is widespread.

Migration-policy nexus

The past three decades have seen a radical transformation in the academic study of international migration. European and North American programmes have funded major academic projects on this area. There has been an influx of academic monographs, general books and journals that have focused on the ever-changing dynamics of migration. This expansion of scholarship and research on migration is not a fluke; it is closely linked to current political and policy preoccupations with economic migration. The research focuses on a range of questions, including new patterns of migration flows, asylum and refuge, multiculturalism, religious and cultural diversity, identity formation among migrant communities and the impact of migration on economic and social development, migration and work, inter-ethnic relations, generational change among migrant communities and the governance of migration. Political and social debates about all these topics have shaped both the scope of research and some of the specific academic preoccupations.

DOI: 10.1057/9781137451187.0008

Migration policy is a matter that has generated much controversy and debate both on the national and international levels. While international migration can be a powerful factor in improving peoples' living conditions, the lack of a comprehensive and coherent policy undermines the potential of this important area (Alonso, 2011). In fact, migrants' living and labour conditions are characterized by being low quality with a high degree of insecurity; this is accompanied by a general lack of knowledge about their rights. Though the terms of employment may be better than in some of their home countries, the migrants' working conditions fall far below those who are citizens in the host countries (ILO, 2005).

It is a basic premise that a state's first responsibility is to protect its citizens. This is true whether the citizens are on foreign or domestic lands. However, labour sending countries seem to be far more preoccupied by promoting overseas employment as an engine for economic development, than in protecting their workers' rights. Though there has been some recent progress made, Bangladesh still faces many challenges regarding their peoples' migration, and any further migration policy adjustments should first address the protection gaps and insecurity of their migrant population.

Policies in South Asia

South Asia has been a central zone of migratory movements throughout history; it has shaped the contemporary demographic landscape. Today South Asian migrants make up a sizeable proportion of the world's migration flows. Afghanistan, Bangladesh and Sri Lanka are three major sources of emigration in the region (Ullah, 2012a).

In the era of globalization human migration has evolved into a dynamic force with people crossing borders at will every day. Some are lucky, and make the trip to their destination countries, many are not. Stories emerge of entire boatfuls of economic migrants perishing in the open seas, while others face the dangerous elements that exist on the world's most porous borders. Despite the risk, movement occurs at continental, regional and local levels. South Asian international migration is not a phenomenon of the modern era, but in the last 50 years thirty million people have crossed the region's border either to meet their basic needs or to escape prosecution.

DOI: 10.1057/9781137451187.0008

International economic migration is a key political issue in the developing world; not only does it directly contribute to the national GDP but the export of labour also eases tension in the labour market at home. South Asian governments promote migration because of the measurable benefits through employment, lessening poverty, and the foreign currency earnings that are remitted back home (Ahn, 2005). They prioritize out-migration through institutional recruitment support and policy implementations that encourage overseas employment. The policies may promote migration, but they do little to protect the rights and interests of migrant workers and their families. Given the dynamics of today's migration, policy intervention should be targeted to not only harness the economic benefits of migration but also reduce the exploitation of migrants.

Most South Asian developing countries face a dilemma between the promotion of overseas employment, and the protection of their labourers working abroad. The focus is on emigration policies with little attention to immigration policies as most of these countries are the countries of emigration. In these countries, migration issues are directed towards development planning or poverty reduction policies. Therefore, the majority of Asian developing countries prioritize overseas employment policies, and their related institutions such as remittance and employment (Rahman and Ullah, 2012).

Most emigration policies in South Asian countries were triggered by the 1970s oil boom in the Middle East. Ministries and their related institutions are responsible for the recruitment and required training for the aspirant migrants. For example, in the Philippines there is a mandatory pre-departure orientation seminar (PDOS) which gives information on documents needed, working rights and obligations, travel tips, remitting earnings mechanisms, and health and occupational safety matters (IOM, 2003). However, the role of state and interstate cooperation is very limited among South Asian countries.

Workers must be protected from the dehumanizing treatment they are subjected to while employed abroad; the importance of formulating comprehensive migration policies cannot be ignored. Most countries have a migration policy at the embryonic stage that have no real influence on procedures and norms in the receiving countries. Migration policies vary from country to country, because each state prioritizes objectives that are the most important to its interests. However, in order for migration to become a powerful driving force promoting aggregated

DOI: 10.1057/9781137451187.0008

TABLE 4.4 *Institutional and legal framework for emigration (Bangladesh)*

Country	Responsible agency	Legislation
Bangladesh	Bureau of Manpower, Employment and Training (BMET); Bangladesh Overseas Employment and Services Ltd. (BOESL)	Emigration Ordinance 1982 Bangladesh Entry Control Act 1952 Bangladesh Passport Act 1920 Bangladesh Immigration Act Bangladesh Citizenship Order 1972 Bangladesh Citizenship (Temporary) Rules 1978 Bangladesh Passport Order 1973 Bangladesh Passport Rules 1974 Passport Rules 1955 The Passport (Offence) Act 1952 The Emigration Ordinance 1982 The Foreigners Act 1946 The Foreigners Order 1951 The Registration of Foreigners Act 1939 The Registration of Foreigners Rules 1966 The Bangladesh Control of Entry Act 1952 Registration of Foreigners (Exemption) Order 1966 The Emigration Rules 2002 Recruiting Agent's Conduct and License Rules 2002 Women and Children Repression Prevention Act, 2000 (Act no. 8 of 2000)
India	Ministry of Labor and its Protector General Emigrants; Ministry of External Affairs; Ministry of Home Affairs	Emigration Act 1986
Indonesia	Minister of Manpower and Transmigration; Coordinating Board of Indonesian Overseas Employment	Ministerial Decree KEP 104 A/MEN 2002
Pakistan	Bureau of Emigration and Overseas Employment; Overseas Employment Corporation	Emigration Ordinance and Emigration Rules 1979
The Philippines	Department of Labour and Employment (DOLE); Philippine Overseas Employment Administration (POEA) Overseas Workers Welfare Administration (OWWA)	Republic Act 8042 1974 Migrant Workers and Overseas Filipinos Act 1995
Sri Lanka	Sri Lanka Bureau Of Foreign Employment; Ministry Of Foreign Affairs; Ministry Of Interior; Ministry Of Women's Affairs; Ministry Of Vocational Training	Sri Lanka Bureau of Employment Act No 21 of 1985
Thailand	Ministry of Labor and Social Welfare; Overseas Employment Administration	Recruitment and Job Seekers Protection Act BE 2528 (AD: 1985)

Source: Adapted from International Organization for Migration (2010).

DOI: 10.1057/9781137451187.0008

efficiency and welfare to the whole of the international system, migration policies must be based on reciprocal understanding and gains (Alonso, 2011). Three basic objectives could be considered for comprehensive migration policy (Wickramasekara, 2009): good governance, protection and empowerment of migrant workers, and promotion of migration's development benefits.

Migration policies of Bangladesh

Bangladesh is one of the most densely populated countries in the world, accommodating around 160 million people in a mere 144,000 square kilometres. The youth population (15–34 years) constitutes nearly one-third of the total population. Only a small proportion of sizable population is employed. So the remaining young people have few prospects at home, and thus choose to migrate. These factors have led to almost four decades of Bangladesh's outward flow of international migration (Sikder, 2008). Bangladesh is one of the leading manpower-sending countries in this region: around seven million Bangladeshis are currently working in 132 countries. In Bangladesh, migration to a foreign country is seen as a prestigious job.

Legislation

Siddiqui (2005) identified the three legal and regulatory frameworks that relate to migration within Bangladesh: international instruments; the domestic laws of both the destination countries and Bangladesh; and the bilateral agreements between Bangladesh and the receiving ends.

Bangladesh has several relevant migration laws, which focus on the two main elements of migration that are within the mandate and scope of the government, maximization of labour migration and the protection and welfare of the migrants working abroad. Recently efforts have been made to consolidate the laws, and to revise and update them. Despite these significant advances, especially in the last decade, progress is needed in setting up a comprehensive legal framework that clearly defines responsibilities and policy implications to structure and implement reforms that more efficiently manage the migration process (Siddiqui, 2003).

DOI: 10.1057/9781137451187.0008

TABLE 4.5   *Existing policies, acts and ordinances related to migration (Bangladesh)*

Name	Category
Expatriates' Welfare Bank (2010 number – 55 Rules)	Act
Special Facilities for Remittance Sender Non-resident Bangladeshi 2008	Policy
Foreign Employment Policy 2006	Policy
CIP Selection Policy 2006	Policy
Recruitment Agency and License 2002	Policy
Wage Earners Welfare Fund Rules 2002	Rule
Emigration Ordinance 1982	Ordinance
Bangladesh Entry Control Act 1952	Act
Bangladesh Passport Act 1920	Act
Bangladesh Immigration Act	Act

Source: Adapted from Ministry of Expatriates' Welfare and Overseas Employment (MEWOE) (2011).

Labour migration can be regulated by bilateral agreements that come in different forms: a formal bilateral agreement and Memorandums of Understanding (MoU). The MoUs are preferred by the recipient countries because they are less formal and non-binding, and thus they are easier to negotiate, implement and modify (Siddiqui, 2005). In the past, Bangladesh signed labour agreements with Iraq, Libya, Qatar and Malaysia. Currently, a Bilateral Technical Cooperation Agreement on Manpower has been concluded with the Government of Kuwait in 2000. However, it is under review and still has not been ratified by Kuwait (UNB, 2010). Bilateral agreements are also underway between Malaysia and Bangladesh, which would replace the 2003 MoU signed by both countries. It should be noticed that bilateral labour agreements are an exception rather than the rule in Asia (Go, 2007). Due to the reluctance on the part of recipient countries to sign these agreements, alternative mechanisms have been developed by the migrant-sending countries to protect their workers abroad (Kader, 2008).

Evolution of migration policy

There are provisions regarding the entry, regulation, and control of foreigners such as the Emigration Ordinance of 1982, and the colonial era Export and Imports (Control) Act of 1950, the Passport Act of 1920, and the Immigration Act. No legal policy framework exists to deal

DOI: 10.1057/9781137451187.0008

with international migration from the country. Despite the constant debates and discourse, the Government of Bangladesh does not have any comprehensive policy on labour migration for overseas employment, whether for skilled or unskilled workers. However, several phases can be identified in the policy evolution; each are defined by dramatic changes in the economic, social and political context of governance.

Pre-independence period (before 1971)

During the British colonial period, human migration within the Indian subcontinent was geographically limited, but historians know that the area that would become Bangladesh was a prosperous destination region. During this period the British colonial policies had an adverse impact on Bengal's industrial sector, devastating the cotton industry, and resulting in a mass migration of people from this region to Assam. Human movement to the other parts of the globe, especially to Europe, also dates back to the colonial era of Bangladesh (Sikder, 2008).

During the pre-independence period, migration was dominated by people seeking permanent settlement. In this epoch a large number of Bangladeshi Hindu families migrated to India because of war. In the early 20th century many inhabitants of this region also moved to Western Europe, North America and Southeast Asia in search of permanent settlement; they contributed to the evolution of the Bengali community overseas (World Bank, 1981; Gardner, 1995; Knights, 1996). Before independence, there was no legislation regarding emigration other than the Emigration Act of 1922. This was because of the colonial government's lack of political will. The perception that overseas workers contribute to a country's development is a relatively new paradigm. There was also virtually no concern for the well-being of national workers. Authorities had very little political awareness or willingness to concern the legislature with matters of migration.

Bangladesh period (1972 and onward)

After 1971 the labour migration was mainly temporary, although a large number of Bangladeshi from higher and middle-income classes continued to migrate to developed countries for permanent settlement (Ullah, 2012; Rahman and Ullah, 2012). In the oil and economic boom of the mid-1970s Bangladeshi workers began to migrate in their droves to the oil-rich Gulf States. This trend ended in the late 1980s, which led

DOI: 10.1057/9781137451187.0008

to the return of many unskilled migrant labourers (Arnold and Shah, 1986). The 1990–1991 Gulf War, and later the US 2003 invasion of Iraq, have seriously disrupted labour markets in the region, leaving many Bangladeshi migrants without work. As the labour flows to the Middle East cooled, Bangladeshi workers found a new market for their skills, and East and Southeast Asian countries became major destinations for labourers. The Asian Tiger boom served as a magnet for the unskilled and semi-skilled workforce seeking temporary employment. Singapore, Malaysia, South Korea, Brunei, Hong Kong and Thailand are still the major labour-receiving countries in this region, although it is Japan that hosts the largest number of documented and undocumented Bangladeshi migrants in the region (Khondker and Rahman, 2009).

Bangladesh has shifted its policy paradigm from one of interventionist to developmental. The 1982 Emigration Ordinance, which replaced the earlier 1922 Emigration Act, was mainly designed to issue licences to recruiting agencies, and ensure the protection of unskilled and semiskilled workers' rights (Rahman and Ullah, 2012). Under the ordinance, the registration of all recruiting agents is required by law. There are also provisions to check irregularities in workers' recruitment, and to appoint labour attachés to look after the workers and punish unlawful emigrants. In this the government's role has been perceived as that of a facilitator, aiding Bangladeshis in the process of finding gainful employment.

According to the constitution, (section 20 and 40), the People's Republic of Bangladesh is committed to assisting human resource development, and providing employment for its citizens. As with any young large population, there are more people than jobs. This pressures officials into actively seeking ways to promote growth through international emigration, as a way of relieving the demand for jobs at home. The Bureau of Manpower, Employment and Training was created in 1976 to facilitate labour migration. It wasn't until 2001 that the Ministry of Expatriates' Welfare and Overseas Employment was established in order to expedite economic development through migrants' remittances.

Protection framework for migrants

A Protection Project of the John Hopkins University formulated a 100-point best strategies for protecting migrant workers. We, in light of the best practices, offer a region-specific framework.

DOI: 10.1057/9781137451187.0008

TABLE 4.6 *Best practices for protecting migrants in the MENA and Central and Sub-Saharan Africa*

Middle East and North Africa	Central and Sub-Saharan Africa
Providing educational, social and legal services for migrant workers	Providing affordable health care for domestic workers
Raising awareness through online social media	Extending social security coverage to migrant workers and their families
Empowering women and children through legal reform and lobbying efforts	Researching options for empowering and protecting domestic workers
Supporting migrant workers through special government agencies	Educating children and mobilizing actors to improve and expand education infrastructures
Providing consultation and legal aid to migrant workers	Working to defend domestic workers and fight against domestic child labour
Providing legal aid, consultation and training to promote migrants' rights	Organizing domestic workers to defend their rights
Facilitating social initiatives for the integration of migrant workers	Using the union to empower domestic workers and enlighten employers
Improving labour law coverage for domestic workers	Promoting the regulation of domestic work
Using MOUs to strengthen bilateral commitment to ethical labour standards	Tackling exploitative practices and violations of children's rights
Training law students on migrants' rights	Using surveys to ignite activism of domestic workers
Promoting regional capacities to combat trafficking	Protecting the rights of vulnerable urban girls
Utilizing public campaign ads to galvanize support for anti-violence legislation	Building the capacities of regional institutions and national governments in the area of labour migration
Creating e-learning courses to reduce workers' risk of exploitation	Seeking to combat unfair instances of deportations and refoulement
	Promoting migrant workers' rights in the workplace through research, training and advocacy
	Providing legal assistance and advocating for migrant workers' rights
	Deterring illegal migration by fostering legal and safe migration
	Securing miners' wages by instituting a deferred wages deposit programme

Source: The Protection Project, 2013 (with permission).

TABLE 4.7 *Best practices for protecting migrants in Asia and the Pacific*

South East Asia, South Asia and the Pacific	Central and East Asia
Establishing a "safe migration" programme for migrant workers	Passing national legislation to fight human trafficking
Identifying areas for cooperative anti-trafficking efforts	Developing pre-departure training for migrant workers
Preventing potential trafficking abuses among migrant workers	Using legal education to prevent human trafficking
Launching a large-scale anti-trafficking campaign	Designing a regional anti-trafficking response.
Providing care for children of migrant construction workers	Enabling safe border crossings through increased bilateral cooperation
Holding employers accountable for the abuse of migrant workers	Preventing labour exploitation in supply chains
Sponsoring workshops to enhance international cooperation	Encouraging safe migration using the railway network
Reducing exploitation of migrants during the recruitment phase	Offering supportive services for migrant workers
Helping migrants gain English fluency to prevent unfair labour practices	Starting a government-run registry for citizens seeking employment abroad
Lobbying for more government action to combat violations of migrants' rights	Conducting a comprehensive anti-trafficking programme
Facilitating information exchange regarding migrant smuggling	Adopting anti-trafficking standards in national trade union policies
Crafting migrant-protective policies based on international standards	
Ending forced labour in supply chains	
Promoting ethical standards for media coverage of human trafficking	
Designing regional solutions to protect migrant workers' rights	
Providing data and reports on trafficking issues	
Improving policymaking through active case analysis	
Empowering migrant workers to access health services	

Source: The Protection Project, 2013 (with permission).

DOI: 10.1057/9781137451187.0008

TABLE 4.8 *Best practices for protecting migrants in the Americas*

Central and South America	North America
Providing migrant protection along the borders of conflict affected states	Offering services to domestic workers
Raising awareness against forced labour in the logging industry	Providing protection for children of migrants
Designing a government action plan to combat slave labour	Establishing laws at the state level that protect domestic workers
Protecting domestic workers through training and legal counsel	Advocating for the rights of female domestic workers
Contracting with the US Department of Labor to protect migrant workers	Providing assistance for migrants in native languages
Conducting health research to influence migrant health policy	Maintaining media campaigns to highlight services for domestic workers
Supporting migrant workers through education and advocacy	Constructing adequate housing for migrant workers
Signing and ratifying international agreements to protect domestic workers	Ensuring educational opportunities for children of seasonal workers
Protecting the rights of domestic workers through legislation	Educating migrant workers on their rights prior to departure from their home country
Raising awareness on domestic workers' struggles through scholarly publications	Maintaining a list of individuals or corporations whose authorization to operate has been revoked
Exercising collective bargaining to promote the rights of domestic workers	Providing health services to migrant farm workers
	Resolving wage theft issues
Protecting the rights of domestic workers through tripartite consultation	Signing bilateral agreements to protect the welfare of migrants
Using legislation to guarantee equal rights for domestic workers	Building emergency water stations for those who cross desert terrain
	Partnering with international organizations to facilitate safe migration

Source: The Protection Project, 2013 (with permission).

DOI: 10.1057/9781137451187.0008

TABLE 4.9   *Best practices for protecting migrants in Europe*

Europe and Eurasia
Using international best practices to design national policies
Unionizing to push for amendments to existing labour laws
Developing universal service employment checks for domestic workers
Providing trafficked persons with residency permits
Opening counselling centres for migrant workers
Ensuring fair wages and working conditions for migrant workers
Educating migrant workers about unionization
Raising awareness of labour rights for potential migrant workers
Championing justice for the undocumented migrant
Designing action plans to encourage robust defence of migrants' rights
Implementing measures to ensure victims' access to compensation
Developing a national referral mechanism
Publicizing a nationwide anti-trafficking helpline

Source: The Protection Project, 2013 (with permission).

Factors shaping migration policy

The role of the state as an active actor in making and shaping migration policies is widely accepted. How a country shapes its emigration policy is closely linked to their history, demography and politics (Yamanaka and Piper, 2005).

Migratory processes are regulated by a wide range of conditions in both the sending and receiving countries, and in the relationships between them. In order to understand any kind of migratory flow it is necessary to analyse all aspects of the societies involved. Factors driving the migration process are numerous, and it is not possible to discuss all of the factors involved with the formulation of migration policy. In countries like Bangladesh, although international labour migration provides the fuel for the economy, migrant remittance as a policy issue received very little attention for many years.

Social dynamics of the migratory process

Migration is primarily determined by market factors (Castles, 2004). Although it is taken for granted that it is the state that intervenes to control migration, Castles argues that the economic and bureaucratic structures are instrumental in shaping migration policy. From the economic school

DOI: 10.1057/9781137451187.0008

of thought, migration ought to be analysed with neoclassical cost-benefit calculations that are indicated by the market behaviour of migration flows. People move if the migration maximizes their individual utility, and end the course of migration when the cost-benefit equation changes. Both internal and international migrations are motivated by the pulls of wage differentials. Within the scope of the bureaucratic process, proponents endorse regulations to categorize migrants and to control their admission and residence. A balanced approach implements bureaucratic controls and protections that recognize and even capitalize upon the economic forces at work.

Contention is sure to arise at the point of intersection between migration policies, economic pull and family needs. Some governments find that their national economy is fuelled by the remittance sent by labourers working abroad and therefore actively encourage emigration. Bangladesh stabilizes their foreign reserve with the remittances sent from abroad, so it is clearly in their interest to develop migration policies that streamline and secure the emigration process. These are not the only forces exerting pressure on Bangladesh's migration policy – special interest groups, politicians, civil society and even media are key components to drafting new socially acceptable legislature. However, the important question remains whether these policies are concerned about the protection of migrant workers.

Migration Governance

Migration governance has come to the forefront because of its enormous significance in the global migration trajectory. Governance is a broad concept that encompasses many different aspects of state and non-state rule; the scope of the term is so vast, it is difficult to delve into it in greater depth within the scope of this chapter. Nevertheless, some key ideas can be identified when dealing with international migration. From the labor migration perspective, the governance refers to the policies that respect fundamental human rights of migrants and the appropriate legislation that supports such policies and the effective administrative machinery for implementing these policies (ILO, 2005).

Governance of international issues is agreed upon by nation-states that see advantages in creating rules and norms, and the institutions to ensure that they are followed. For example, the World Trade Organization (WTO) establishes rules for international trade in goods and services, establishing mechanisms to resolve trade disputes; the

DOI: 10.1057/9781137451187.0008

International Labor Organization (ILO) promotes policies that protect workers. Nation-states may delegate a part of their national sovereignty to international institutions such as the WTO and ILO, agreeing to incorporate the agreements or conventions they establish into national laws governing trade and worker rights.

International organizations, such as the WTO and ILO, set rules for the behaviour of nation-states, while the International Organization for Migration (IOM) provides services to states. The IOM began as an inter-governmental organization that moved refugees and displaced persons to new homes at the end of the Second World War. It has evolved into an organization in 118 countries that aims to improve migration management by providing services and advice to governments. The IOM acts as secretary for a variety of ad hoc efforts to improve migration management, including the Berne Initiative that led to the International Agenda for Migration Management, a "non-binding reference system and policy framework" (IOM, 2009; Ullah, 2009). However, it is clear that the IOM is overburdened with the responsibility of such a huge population around the world. The contemporary world has been witnessing the newly uprooting of millions of people over the last 4 years, especially in the Arab regions. In many cases, for example, during the political turmoil in Libya, a huge number of migrant population were trapped in Salloum area (Egypt-Libya border), many were injured and a few others were killed. A condition like humanitarian crisis took place. The government seemed to have surrendered to the situation and the IOM had to take the responsibility to take care of the situation.

Recently, migration has risen to the top of many national and regional policy agendas (Ullah, 2008), because of its consistent and persistent significance in economic growth and development across the world. The role of public administration is vital for efficient migration in both the public and private sphere. International and transnational organizations dealing with migration have recently underscored the importance of migration governance. Due to the lack of governance, migration management quality varies considerably across the globe. The subsequent outcome is poor migration governance that results in negligence towards migrants and their families (Ullah, 2009).

The migration governance can be analysed on different levels of political action, which are not mutually exclusive, but are actually complementary. Firstly, governance can be studied from a supranational perspective, taking into account the international system and the need for a global governance. From this perspective, the main subject of study rests on the need to provide joint and global responses to the migration phenomenon, since it represents a global challenge. Alonso (2011) argues that properly regulated migration can be a source of profit shared by all individuals, societies and the governments involved. Therefore, it should be perceived as a win-win situation. This argument implies that both the countries of origin and destination should strive to mitigate the negative impacts. In order to reach this objective, sending countries should be aware that only a strong, coherent and solvent development policy can maximize positive effects. The host countries should be aware that they meet their development needs through migrant workers who require legal protection, recognition of social rights and non-discriminatory conditions (Alonso, 2011). These ideals have not yet been actualized, inconsistencies within migration policy can be found in both sending and receiving countries.

A labour sending country, such as Bangladesh, requires good govern ance to coordinate the action of many actors inside the country, and to maximize the efficiency of policy administration in order to minimize the incidence of fatalities. Agencies may find that they are duplicating each other's functions, and undermining each other's effectiveness. Therefore, a clear leadership, that establishes priorities and decides on jurisdictions, seems to be essential in ensuring good governance (ILO, 2005).

The public and private recruiting agents in Bangladesh include BOESL (Bangladesh Overseas Employment Services Limited), the BMET and personal initiatives. The government monitors and supervises the overall recruitment process through the BMET. Available data on migration confirm that in 1976, 87 per cent of temporary migrants migrated through governmental sources, while 4.66 and 8.60 per cent migrated through recruiting agents and individual initiatives respectively. In 1999, 58.6 per cent of migrants migrated through individual initiatives, while 41.26 per cent migrated through recruitment agents (IOM, 2002). This means that personal initiatives and migrant networks are increasingly becoming the prime source of channelling overseas migration, and may mean that the government's role is getting weaker in sending migrants.

In 2011 Bangladesh hosted the fourth round of the "Colombo Process", a series of regional consultative meetings on Asian contractual migrant workers. Under the theme "Migration with Dignity", delegates from 11 Asian countries discussed strategies to improve coordination, optimize benefits from migration, and prevent abuses at home and abroad.

DOI: 10.1057/9781137451187.0008

Several labour-receiving countries from Asia and the Middle East attended as observers. However, no improvement in Bangladesh has appeared visible yet.

The fact is that no global migration governance regime exists to regulate migration and oversee how states discharge their responsibilities. As a result, some countries, mostly the Gulf States, do not bother with welfare issues of the migrant workers (Hansen, 2010). It was against this background that the Regional Consultative Process (RCP) on Migration emerged in the late 1990s in many regions of the world. In Asia the following RCPs can be identified:

▸ Inter-governmental Asia-Pacific Consultations on Refugees, Displaced Persons, and Migrants (APC), first held in 1996.
▸ The Bali Process on People Smuggling, Trafficking in Persons and Related Transnational Crime, first held in 2002.
▸ The Ministerial Consultations on Overseas Employment and Contractual Labor for Countries of Origin and Destination in Asia (Abu Dhabi Dialogue), 2008. This commitment is embodied in the Abu Dhabi declaration, which aims to maximize the benefits of temporary labour migration, both in countries of origin and destination, as well as labour migrants themselves.
▸ Ministerial Consultation on Overseas Employment and Contractual Labor for Countries of Origin in Asia (Colombo Process), first held in 2003.

Bangladesh has attended all of the RCP's as a state participant. However, it can be said that the more active role Bangladesh has played was under the Colombo Process. Afghanistan, China, India, Indonesia, Nepal, Pakistan, the Philippines, Sri Lanka, Thailand, Viet Nam and Bangladesh all participated in the talks. Another eight countries attended as observers; Bahrain, Italy, Kuwait, Malaysia, Qatar, Republic of Korea, Saudi Arabia and the United Arab Emirates. The three thematic focal points of the Colombo Process: the first provision was the protection and provision of services to overseas temporary contractual workers. In particular, it focused on protecting contractual workers from abusive practices during recruitment and employment, and providing them appropriate services in terms of pre-departure information, orientation and welfare provisions. The second element of the meeting focused on optimizing the benefits of organized labour mobility. This includes the development of new overseas employment markets, increasing

DOI: 10.1057/9781137451187.0008

remittance flows through formal channels and enhancing the development impact of remittances. The final element was to focus on capacity building, data collection and inter-state cooperation. They aimed to increase institutional capacity building and the information exchange in order to meet labour mobility challenges, which increased cooperation with destination countries in the protection of overseas temporary contractual workers, and the access to labour markets; and enhancing cooperation among countries of origin (IOM, 2011). However, violent torture and death as a result were not found as important issues on the agenda.

Migrant Forum in Asia represents over 200 groups in 17 countries across the region. Human Rights Watch is an international non-governmental organization that investigates human rights violations in some 90 countries, and has published several reports on the abuse of migrant domestic and construction workers in Asia and the Middle East. Coordination of Action Research on AIDS and Mobility (CARAM) Asia is a regional network of non-governmental and community-based organizations focused on migration and health with members in Asia and the Middle East (CARAM Asia, 2011).

The Migrant Forum in Asia, Human Rights Watch, and CARAM Asia welcome the reinvigoration of the Colombo Process, a series of regional consultative meetings on Asian labour migration. Since the first ministerial consultations in 2003 there have been another three ministerial consultations: in Manila 2004, Bali 2005 and Dhaka 2011. The last meeting was held in Dhaka in April 2011, and may represent a new impetus to the process after years of stagnation. New recommendations were made at the Dhaka declaration to strengthen five main areas of migration governance: Promoting Rights; Welfare and Dignity: Services and Capacity Building; Emergency Response and Emerging Issues; Enhanced Dialogue and Cooperation.

These new cooperative commitments must be considered with caution; after more than 9 years there has yet to be any concrete action in relation to the protection of the workers' rights (Palma, 2011). This is probably due to the nature of the informal and non-binding process.

Institutional framework

There are a number of ministries and departments that are involved in the process of international migration. Their functions overlap and there is a lack of coordination between and among the departments.

DOI: 10.1057/9781137451187.0008

Ministry of Expatriates' Welfare and Overseas Employment (MEWOE)

The MEWOE was established in 2001 and it is the leading ministry in charge of migration issues. The main objective of the ministry is to ensure the welfare of the expatriate workers, and the enhancement of overseas employment. The Ministry is dedicated to improving the flow of remittances, providing an equal opportunity for all people seeking overseas employment, and ensuring the overall welfare of migrant workers (MEWOE, 2011). It deals with the process of recruitment and placement, suggests legislation and implements laws regarding migrant workers (IOM, 2002). The BMET is the only active unit. Earlier this division was under the direct control of the Ministry of Labor and Welfare, but now it is operating under the Ministry of Expatriate Welfare and Overseas Employment (MEWOE, 2011).

The Bureau of Manpower, Employment and Training was established in 1976 by the Government of the People's Republic of Bangladesh, as an attached department of the then Ministry of Manpower Development and Social Welfare, with the specific purpose of meeting the manpower requirements of the country and also for the export of manpower. The BMET is engaged in all planning and implementation of the strategies for the proper utilization the nation's manpower. The BMET is currently under administrative control of the Ministry of Expatriates' Welfare and Overseas Employment. The Bureau is performing all functions relating to the migration process including licensing of recruiting agents. The BMET provides training in employable trades through 14 Technical Training Centers (TTC), one Institute of Marine Technology and three Apprenticeship Training Offices across the country. The BMET conducts training activities, and provides institution-based vocational training in different employable trades and offers pre-departure orientation.

The BMET is the most vital government agency that deals with manpower export. It is in charge of processing foreign demands for the recruitment of Bangladeshi workers, as well as controlling and regulating emigration clearance for workers recruited for overseas employment. It is also responsible for the registration of unemployed persons, and the referral to vacancy positions and regulations of the private Recruiting Agents.

Bangladesh Overseas Employment and Services Limited (BOESL)

The key functions of the BOESL are: promoting Bangladesh as a reliable source of quality manpower by means of regular publicity and

DOI: 10.1057/9781137451187.0008

advertisement in the international print media; undertaking effective employment promotion campaigns in the countries requiring manpower, to secure employment offers from potential receiving countries; and arranging employment tests, medical tests, tickets and other facilities for persons selected for foreign employment.

The Bangladesh Overseas Employment and Services Limited is intended to be a model institution in managing the manpower and operating in competition with about 700 private recruiting agencies in the country. It is dedicated to earning foreign exchange by way of exporting skilled and unskilled manpower and ensuring supply of quality workers within the shortest time span. The main purpose of creating this company was to provide honest, efficient and quick services to the valued foreign employers in the matter of deployment of manpower development.

Bangladesh Association of International Recruiting Agencies (BAIRA)

The BAIRA was also established in 1984 as a national level association for cooperation and the welfare of the migrant workforce. It has approximately 700 member agencies in collaboration with and support from the Government of Bangladesh. The BAIRA generally promotes and protects the rights and interests of the members of the association; interacts with the government, foreign missions and employers to facilitate the migration process of Bangladeshi workers; and explores job opportunities, arranges for training and facilitate the migration process and promote the welfare of Bangladeshi workers abroad and increase the flow of remittances to Bangladesh.

With a view to ensuring the financial security of migrating workers, the BAIRA has already undertaken two insurance programmes: one for the workers before their departure and the other for their families left behind, through the BAIRA Life Insurance Company Limited, an affiliated organ of the BAIRA. In addition, the BAIRA set up a bank to ensure quick and easy monetary transactions for members of the BAIRA and the migrating workers. They are currently launching a sophisticated and highly technical vocational training centre to train youths in Information Technology. The BAIRA is also setting up a medical testing centre, with advanced facilities in place to ensure that all migrant workers receive a comprehensive check-up. It arranges short-term foreign language courses and briefing sessions before allowing the selected workers to emigrate (BAIRA, 2013).

DOI: 10.1057/9781137451187.0008

Bangladesh Missions

Presently, there are 60 Bangladesh missions abroad, comprising 46 embassies/high commissions, two permanent representative missions to the United Nations, two deputy high commissions, with the remainder being sub-delegations (Rashid, 2011). These missions in the host countries are supposed to assist migrant populations in times of need. The main functions offered by the delegations abroad include: hearing complaints lodged by migrant workers, providing legal assistance, taking up the matter with employers, and repatriating stranded or deceased migrant workers. However, there are widespread complaints about severe inefficiencies and corruption of the mission officials.

Establishment of Probashi Kalyan Bank

The primary objective of establishing a Probashi Kalyan Bank (Expatriate Welfare Bank) was to lower the travelling cost for potential migrants. To that end, the bank provides loans at a lower interest rate than what was previously available, and sets up a legal and safe way for migrants to send remittances home. The bank began its functions in 2011. It provides potential migrant workers with collateral-free soft loans, and migrant workers already abroad with low-cost and easy remittance transfer services. The bank would offer loans at 9 per cent interest rate, which is said to be much lower than what commercial banks provide. The belief is that this initiative eases the migration process for the potential migrants who used to resort to informal and formal moneylenders to obtain capital to finance their migration, paying an exorbitant rate of interest.

Welfare fund for migrant workers

There is a welfare fund operating in Bangladesh that provides financial assistance for legal support, support for initial sustenance, repatriation of migrant workers from host countries, and legal support to family members in the country of origin. Under this fund, a foreign service was developed in order to host increased numbers of guests from other services, and yet it somehow failed to maintain the sanctity of the service itself. Part of the problem has been that political appointees are being stationed in diplomatic offices.

Though significant progress has been made on migrant issues, the ideal scenario of "Migration with Dignity", however, is still far away. Unethical practices were the major challenge, as unscrupulous middlemen take advantage of the desperation of poor people, wishing to go overseas. The

DOI: 10.1057/9781137451187.0008

Government of Bangladesh seems to have attached emphasis on laying down the regulatory legislation as part of the Emigration Ordinance of 1982. Drafting on an "Overseas Employment Policy" is also under way in this regard (Haque, 2012).

Policy challenges

It is important to consider what policies the governments of both host and sending countries have implemented to reduce, if not stop, incidents of abuse of migrant workers. A common problem that occurs on the policy level is the reluctance of the sending state's government to take adequate action to protect its workers abroad in a timely manner. Different reasons exist. Migrant workers generate large amounts of remittances, which the government, as well as the families of the migrant, cannot afford to lose (*The Economist*, 2011). Governments of sending countries rely on this currency injection, and cannot afford to stop sending migrant workers abroad.

Another reason that explains the poor protection of migrants is the foreign policy of the sending countries. In fact, in order to maintain good relations with the host countries, most sending states prefer to postpone or sideline these "delicate issues". The act of accusing powerful states of violations of labour standards, or even in worse cases violations of human rights, can definitely compromise economic relations. Andrzej Kapiszewski, a Polish sociologist and diplomat, discusses how migration plays a major role in foreign policymaking, elucidating that its effect on foreign policy could be quite significant. Indeed, he considers migration an important foreign policy issue, explaining that migrants inevitably affect the policies in both home and host countries (Kapiszewski, 2006).

It is helpful to observe through a political lens how migrants influence the policies of host and sending states, and the different foreign policy challenges that emerge. Migrants, as Kapiszewski (2006) explains, use the local media to exert influence on their host countries' foreign policy. A second effective way is through personal relationships with top-ranking nationals. When migrants are present in large numbers within a host country, the relationship between the latter and the sending country may become complex. An example given by the same author is that of Qatar's accusation that in 1996 Egypt played a major part in the attempted coup.

DOI: 10.1057/9781137451187.0008

As a result, 700 Egyptian workers lost their jobs in Qatar, especially those who were working for the Ministry of Interior. More Egyptian workers lost their jobs in Qatar in subsequent years, due to Cairo criticizing Doha of forming new relations with Israel (Kapiszewski, 2006).

The magnitude of fatalities endorse the fact that sending and host states are averse to providing adequate protection to migrant workers. Some argue that abuse may stem from social norms, therefore not much could be done at policy level. Rachel Silvey discusses how gender politics affect migrant workers, explaining that the states' inability to extend protection to migrant workers is due to the presence of norms pertaining to class, nationality and gender. Such norms are the determinants of which crimes are tolerated and against whom it is unproblematic to commit them (Silvey, 2004).

Another important reason is the lack of bargaining power. Host and sending countries are often unequal in terms of bargaining power. It is believed that the sending countries are in too weak a position to take any stance (Human Rights Watch, 2008). This should not be the case when migration is believed to be a function of demand and supply. Still, it has been noticed that some important steps have been taken by some sending states to provide protection to their workers abroad. The Philippines, Indonesia, and Nepal are the best examples. In the Philippines, females below the age of 21 are not allowed to migrate on their own (Kapiszewski, 2006). Since August 2011, Indonesia has suspended the migration of Indonesian labourers to Saudi Arabia. Such a moratorium was to remain in effect until the two countries implemented proper protection policies for the workers (*The Wall Street Journal*, 2011). Not all protection measures in place are successfully implemented. The proposed moratorium was to prevent "unskilled" workers from travelling abroad for employment. Activists were against this decision as prohibition from working as foreign domestic workers is not one of the best solutions. Foreign embassies stationed in migrant-receiving countries have been blamed for inadequate responses to the needs of the migrant population. Embassies are found to be reluctant to aid migrant workers because deportation may involve a financial burden, one which embassies do not want to carry.

In order to reach a comprehensive analysis on the fatalities of migrant labourers we need to focus our attention on the policies the sending governments have in place. The failure of sending countries' governments and consulates to provide sufficient protection to the migrant

DOI: 10.1057/9781137451187.0008

workers is evidenced in the research this book is built on. South and Southeast Asian countries also have significant gaps in their protection of workers abroad, which leaves the workers in a vulnerable position (Human Rights Watch, 2010b). The fact is that, except for Sri Lanka, no countries in South Asia have ratified the 1990 International Convention on the Protection of the Rights of All the Migrants Workers and Members of Their Families (Haque, 2012). The policies of the sending countries towards the host states can be crucial in closing this gap. Embassies which prioritize worker's rights over diplomatic relations will probably be more responsive towards abuses and show more initiative to ensure the rights of their compatriots. However, embassy officials from labour-sending countries in the United Arab Emirates explained that an embassy's ability to act is dependent on the existence of contracts. Unless contracts have been signed by the worker, the embassy does not have much leeway to exert their rights (Human Rights Watch, 2009). In Bangladesh, where migrant workers are required to register with the BMET and attend a briefing and training programme prior to departure, studies by OKUP (2009) and Afsar (2009) indicate that many do not attend these training sessions. In fact, many of the participants reported that those training sessions were not useful.

Laws and initiatives by receiving countries

It is important that receiving states have relevant policies and legislations in place to properly deal with the inflow of migrant workers (Shah, 2009). Only a few countries in the Middle East seem to have attempted to take steps towards improving the situation of migrant workers. One such attempt is improving housing conditions, and another, better access to health care. Another important initiative is that of not making it compulsory to work during the hottest hours of the day in summer. There are, however, questions and debates about the enforcement of this policy. Human Rights Watch (2009) reported that the government closed some so-called "labour camps" in which workers used to live, and the Government agreed that 40 per cent of the Emirate's 1,033 labour camps violated minimum health and fire safety standards (Human Rights Watch, 2009).

In August 2009, a new rule was made effective, which granted migrant workers the right to change jobs without their employer's permission.

DOI: 10.1057/9781137451187.0008

This seems, apparently, to be an excellent initiative. However, it makes no difference to the lives of domestic workers because they are excluded from this rule. In the late 1990s, Lebanon modernized their record-keeping system, such as computerizing the names and addresses of all sponsors and foreign workers (Jureidini, 2004). In 2010, some steps were taken against individual employers who were abusive towards their employees. The US Department of State (2010d) reported the following: "On June 24, the first instance penal judge of Jbeil sentenced an employer who physically abused Sri Lankan domestic worker to one month's imprisonment and a fine of 10 million pounds ($6,666). The woman was also barred from sponsoring or employing a domestic worker for 5 years from the date the decision was rendered. In December 2009 a criminal court judge in Batroun sentenced a woman who beat her domestic maid to 15 days in prison, 50,000 pounds ($33) fine, and 10.8 million pounds ($7,200) in compensation. The victim was in the Philippines when the verdict was announced, and the judge refused the employer's request to bring the victim before the court" (US Department of State, 2010, pp. 35–36). However, prosecution against abusive employers does not often take place. In many cases the victims refuse to press charges because they fear further abuse and lack of evidence.

Saudi Arabia's Ministry of Labor established a Migrant Workers' Welfare Department to provide services pertaining to protecting migrant workers' rights. Labour regulations are in place for migrant workers, but do not apply to domestic workers. These regulations include a 48-hour standard work-week at regular pay; a weekly 24-hour rest period (normally on Fridays, although the employer may grant it on another day) based on hours worked. In 2005, Qatar created a new department for the protection of domestic workers to receive complaints and impose penalties (Jarallah, 2009). On 30 May 2014, six dead bodies arrived in Dhaka from Saudi Arabia. In fact nine were killed in a furniture shop fire near Riyadh in Saudi Arabia on 12 May 2014. We are putting this recent development into this context as we are talking about policies in place and progress in the provisions of upholding the rights of the migrant workers. It remains entirely unknown whether these migrants had their salary unpaid.

Human Rights Watch (2010a) studied 114 cases of abusive employers in Lebanon. This study in fact confirmed the widespread belief about the abusive treatment given to domestic workers in Lebanon. Human Rights Watch found that none of the abusive employers faced charges. Saudi

DOI: 10.1057/9781137451187.0008

Arabia is said to be maintaining a database of abusive employers who, in principle, are prohibited from recruiting new foreign workers. This current study confirms that this system is not adequately implemented. It is evident that most Asian sending countries' states have not investigated migrant deaths, as is demanded of them. The poor implementation of policy, weak foreign policy, inefficient officials employed in foreign missions, and absence of bilateral agreement may have contributed to this situation. Considering the comparative advantage of the labour, there is a need for inter-union cooperation between sending and receiving countries. Effective networking, information-sharing and monitoring may help reduce the rate of fatalities of migrants (Ahn, 2005).

Note

1   http://www.time.com/time/specials/2007/article/0,28804,1779365_177936
    6_1779370,00.html

DOI: 10.1057/9781137451187.0008

5

Conclusions and Recommendations

Ullah, AKM Ahsan, Mallik Akram Hossain and Kazi Maruful Islam. *Migration and Worker Fatalities Abroad.* Basingstoke: Palgrave Macmillan, 2015. DOI:10.1057/9781137451187.0009.

DOI: 10.1057/9781137451187.0009

The proliferation of people migrating on a global level, and the attendant financial transfers in the form of remittances, has brought migration to the forefront of the global development debate. However, migration has not had an equal impact on everyone. While for many, migration made huge positive changes in their lives, many others lost more than they gained. Migration losses come in myriad shapes, forms and scales.

A major form of loss is their life. The incidence of death among migrant workers abroad is escalating. Media has been continuously reporting on arrivals from abroad of the dead bodies of Bangadeshi migrant workers. The *Nayadiganta*, 18 July 2009, reports that an average of seven dead bodies of migrants return each day, which is alarming. The majority of causes of death, according to the death certificates, are cardiac arrests, workplace accidents and motor vehicle accidents.

The conventional way of explaining these incidents are that Bangladeshis live and work in precarious conditions, particularly in the Middle East and Southeast Asia and as a result they are exposed to sudden fatalities. This is endorsed by the fact that most of the migrants who died in factory accidents were from the Middle Eastern countries followed by Southeast Asian countries, particularly Malaysia. Whether or not, this happens accidentally or on purpose, this fact stands as a testament to poor workplace safety for migrant workers.

Unskilled migrant workers are routinely subjected to hazardous and risky situations. Even if hazardous conditions do not lead to an immediate danger of death, the stress and hardship these workers suffer is likely to result in a shorter life expectancy. Examples of the insecurities and dangers prevalent in the workplace environment are infectious disease, and chemical and pesticide-related illnesses.

The notion of protection for migrant workers encompasses provisions for repatriation, the return of remittances after death, and caring for the families of those who die abroad. The level of protection of migrant workers depends largely on the policies in place in both origin and destination countries. This book focuses on the lack of protection policies that result in the inhumane treatment of migrant workers, and frequently in their deaths as well. We have discussed the deaths of migrant workers, specifically Bangladeshi migrant workers, in the context of human rights and migrant rights. Both national and international standards have been considered.

In the absence of adequate national and international protection, employment agencies play a significant role in determining the fates

DOI: 10.1057/9781137451187.0009

of migrant workers. Examples of recruitment abuse, forced labour, and trafficking are numerous, and include providing false information regarding work conditions, increasing recruitment fees thus resulting in indebtedness, threatening workers, hiding information about contracts, and being reluctant to provide assistance to migrant domestic workers (Human Rights Watch, 2008). Prior information is provided to help potential migrants mentally prepare for the new environment. Therefore, misinformation and disinformation could be catastrophic for them (Ullah, 2013, 2013a). Employers hold enormous control over employees, particularly in the domestic sector. Human Rights Watch documents numerous cases of forced confinement, deprivation of food, and physical abuse targeting domestic workers; despite such blatant human rights violations, domestic workers in particular are considered beyond the reach of legal protection as they are employed in the "private" sphere rather than in the "public" sphere. Given they have the capacity, we urge Human Rights Watch to conduct a global study on the causes of death of migrant workers.

Female migrants are generally recruited under a 2-year contract. This type of contract has produced lots of controversies in terms of contract conditionality, harassments, and physical, sexual and verbal exploitation. Contract and work conditions must be based on solid legislative standards or else the workers will be left little leverage to negotiate. Many studies have confirmed that South Asian migrants pay larger sums of money than the amount determined by their respective governments to finance their migration. This forces them to either sell their assets or borrow further funds with high interest charges. This in turn means that they have to stay put in the destination countries, despite knowing they will open themselves to abuse.

Following the death of a migrant worker overseas, the immediate priorities for family members include securing the return of the deceased's body, and accessing any earning that were in the possession of the deceased. In this regard, experiences explained to us in this research varied widely. In securing the return of the migrant's body, those left behind in receiving country relied on networks of family members, friends, and employers to raise the funds required to secure transportation cost. In some cases, the IOM extends support. In the case of irregular or undocumented migrant workers, the financial burden of retrieving the deceased's body typically falls on the family, as irregular employees benefit from even fewer legal protections in the case of death than legally employed migrant workers.

DOI: 10.1057/9781137451187.0009

Family members must often resort to collecting money from friends, relatives, and acquaintances in order to raise sufficient funds to secure the return of the body. No one in our study reported to have received any support from their government.

According to interviews with High Commission officials, prior to 2009 the High Commission used welfare funds to fund the return of deceased Bangladeshi migrant workers in the event the families could not afford transportation costs. However, in 2009, the High Commission ceased using welfare funds for the purpose of transportation and offered assistance in providing documentation for the deceased. The legal and regulatory frameworks that govern labour migration from Bangladesh are international instruments, laws of both the countries of destination and Bangladesh, and bilateral agreements between Bangladesh and receiving countries. Bangladesh's migration laws and policy focus on two main objectives: maximizing labour migration, and ensuring the protection and welfare of migrants working abroad. However, despite significant advances in the last decade, more work is needed to establish a comprehensive legal and policy framework that clearly defines respon-sibility for implementing migration policy reform and managing the migration process.

Although both sending and receiving states have opportunities to implement laws and policies aimed at reducing (if not eliminating) the level of abuse of migrant workers, they are also constrained by myriad factors. Sending governments that prioritize the well-being of migrant workers over diplomatic relationships with host states would be more responsive to reports of abuse, and would take the appropriate initiative to ensure the protection of migrant workers. Sending states, however, face some significant challenges. First, sending states may fear that any action taken may result in the diminution of remittances from overseas workers. Given that remittances are a key source of both family income and foreign exchange, governments of sending states are reasonably leery of jeopardizing remittances.

Second, the need to maintain positive relationships between sending and receiving states, which are frequently politically and economically influential, makes sending states reluctant to assert workers' rights when faced with abuse in host states. Andrzej Kapiszewski discussed the role migration plays in foreign policymaking, suggesting that migrants inevitably affect the policies of host countries as well as their own. He argues that migrants use both local media and relationships with host

DOI: 10.1057/9781137451187.0009

country officials to influence foreign policy, and that the presence of large numbers of migrants can inject greater complexity into relationships between sending and receiving states. Kapiszewski (2006) offers the example of Egyptian migrants who played a key role in a 1996 coup attempt in Qatar. As a result of the involvement of Egyptian migrant workers, 700 Egyptians – particularly those working for the Ministry of Interior – lost their jobs in Qatar. Similarly, Human Rights Watch (2008) asserts that an unequal power balance between sending and receiving states manifests itself in inequitable bilateral labour agreements, which favour host states at the expense of migrant workers.

Third, foreign embassies may be reluctant to aid migrant workers because deportation poses a financial burden for embassies and consulates. In the case of Saudi Arabia the Ministry of Social Affairs prohibits migrant domestic workers from working in Jeddah, where numerous consulates are located, making locating and repatriating migrant domestic workers difficult and costly (Human Rights Watch, 2008). Both sending and receiving states are constrained by social norms in their respective jurisdictions. For example, gender politics can have an impact on attempts to protect migrant workers due to the presence of norms related to class, nationality and gender (Silvey, 2004). Domestic workers are particularly vulnerable in this context, as, in addition to being considered of less value due to both their gender and nationality, they are considered to work in the private sphere and thus are frequently excluded from laws and policies aimed at protecting migrant workers. There has been a noticeable tendency for sending countries to place themselves in a subordinate position. They often tend to forget that this is entirely a reciprocal deal. An inherent idea most sending countries hold is that they are at the mercy of the receiving countries, thus weakening the bargaining capacity for the protection of migrants.

However, several sending states have made sincere efforts to protect their overseas workers. As we mentioned earlier in this book, the Philippines restricts females under the age of 21 from migrating on their own (Kapiszewski, 2006), while the Indonesian government put a moratorium in 2011 on the migration of Indonesian labourers to Saudi Arabia until the two countries implemented proper protection policies for Indonesian workers (*The Wall Street Journal*, 2011). We have visited a number of consulates/embassies in different countries (such as Bangladesh embassies; the Philippines embassies, Sri Lankan embassies, Royal Thai embassies etc.) to gain an idea of the services they provide to their population

DOI: 10.1057/9781137451187.0009

living abroad. It is in fact vividly clear that complaints against Bangladesh embassies from overseas workers hold true. Political recruitments to the embassies of Bangladesh are common. Most of them are not trained and diplomat. A common claim from the service seekers is that their primary agenda is to make money. Therefore, providing services to the people become a trivial matter to them. In Bangladesh, migrant workers are required to register with BMET and to attend a brief training programme prior to departure. However, studies by the Ovibashi Karmi Unnayan Program (OKUP) (2009) and Rita Afsar (2009) found very slim attendance in the training programme and those who attended didn't find it very useful. Afsar (2009) also revealed ignorance about the programme, as well as the practice of employment agents who fraudulently obtain attendance certificates on behalf of clients.

In Saudi Arabia most migrants are not aware of their legal rights. The *Kafala* system (sponsor system) is another form of legal system which is applied in Saudi Arabia and represents a visible form of human rights violations. Foreign workers are restricted to working without having a sponsor. Under this system they cannot change their employment without the consent of the employer as their passports are taken away from them.

As evidenced throughout the book, migrant workers are subject to varied inhumane treatments. They are left with little to no choice when they arrive in the host state, and frequently find conditions different from those set out in the contract that they signed before they flew. They are forced to accept whatever discrepancies they find, because they arrive at a point of no return. In many cases, these contracts are written in different languages from the language the migrant workers can read. Confiscation of passports is a common practice among employers and recruiting agencies, and migrant workers are often subjected to forced labour. Lack of health insurance is a critical issue that needs attention at the policy level. Migrant workers have the right to medical care when they need it, and it is in the best interests of the employers to keep their workers healthy, but this is often seen to be an unimportant issue.

Foreign workers are also deprived of their legal rights. Many countries exclude migrant workers from their domestic labour laws. Not allowing them to be members in labour unions and barring them from joining any associations are also a deprivation of their rights. This, in a way, enables employers to exploit migrant workers with impunity. Clearly, the lack of solid and effective legislation regarding the migrant population remains a

DOI: 10.1057/9781137451187.0009

major cause for the reticent attitude of employers. There have been some attempts by sending states to provide protection to their workers abroad, but more needs to be done. Bangladesh, as well as other South Asian countries, would do well to follow the Philippines' precedent regarding migrants' rights and protection.

The role of receiving countries is also crucial in this matter since migrant workers live and work according to the policies and laws of that host country. The long-held mentality that foreign workers take away their jobs should be changed. Otherwise increasing xenophobic treatment and violence will not decline. We would strongly argue that Asian sending countries receive the highest number of bodies of deceased migrant workers from Middle Eastern countries. This fact can be endorsed by a number of recent brutal acts perpetrated upon Sri Lankan domestic helpers in the KSA. A couple of months ago, one migrant was rescued from an employer's house in a critical condition. Doctors removed 24 nails from her body. This level of brutality and inhumanity towards migrant workers is not uncommon in the region. It is not acceptable, and it deserves the attention of all associated governments.

Complacency is common among the host country governments in that they compensate if any fatality occurs. The definition of compensation however has been distorted by compensation practices. The cost of transporting the bodies of dead migrants home should not be considered compensation. Compensation standards have to be streamlined and incorporated into the policy agenda.

On the national and international level, migration policy has been one of the controversial and debatable matters. International migration is supposed to be one of the factors that improves the lives of people. Yet, lack of a comprehensive and coherent policy undermines its potential. Migration today encompasses a host of issues involving human rights, public health, international security, discrimination and xenophobia. It is imperative to integrate a holistic approach when dealing with these issues.

In terms of the legal procedural framework, Bangladesh deals with migratory matters through a series of international instruments that serve as policy frameworks involving the laws of both the countries of destination and Bangladesh, and bilateral agreements between Bangladesh and the receiving countries. Bangladesh is a signatory state to different bilateral agreements with Iraq, Qatar and Malaysia, but the critical issue is that bilateral agreements are not the norm in Asia, which

DOI: 10.1057/9781137451187.0009

makes it inactive in terms of implementation. With regard to internal policies, the Government of Bangladesh does not have any comprehensive policy on labour migration or overseas employment.

Migration policy is shaped by economics and bureaucratic systems. But who do we think are the writers of the regulations? The answer is that the writers are the big and main players in the recruitment business who have influence in bureaucracy. In addition, three important factors have a major influence in shaping the migration policy: the social dynamics of the migratory process; the effects of globalization, transnationalization and North–South relationships; and the political systems (IOM, 2010). The family can provide financial and social support through links that propagate a migration strategy and facilitate the migration process.

Behaviours of countries across the world are governed by the international system represented by international organizations such as the IOM, WTO and ILO. However, lack of governance results in diversity in migration management among states. This role of governance applies to both sending and receiving states. Both of them should enhance positive impacts of migration and mitigate negative ones. On an international level, although there are laws related to the migration process and migrant workers, what is lacking is a regime that governs laws and their implementation in terms of the protection of migrant workers. There is a Regional Consultative Process that has emerged in many regions. In Asia particularly, this can be seen in international non-governmental organizations, regional network of grassroots organizations, associations, trade unions, faith-based groups and individual advocates across Asia. Nevertheless, there has been no concrete action in relation to the protection of workers' rights. This is due to the informal nature of the process.

Declarations and good policy intentions do not automatically produce good governance, or positively impact workers' conditions. Effective solutions, such as the establishment of funding institutions, must be implemented with a specific political will by the government. Also, the government should increase its budget to apply and implement the drafted policies. Accountability is another important element for the effectiveness of migrant protection since the one who is in charge of managing and implementing the migration sector has to be accountable for the strategies and policies undertaken.

Poor recruitment practices are at the root of migration issues. When this process is run by private agencies abuses occur in most of the cases.

DOI: 10.1057/9781137451187.0009

Within these agencies there is little oversight, which results in illegalities and migrants being cheated. Therefore, the role of the government should be reinforced in controlling the implementation of the migration process. This can be done through a specific organization that deals with migrant worker matters such as documents, training, culture, obligations, regulations, rights, duties and adaptation for workers in the host state. The role of embassies should be enhanced to support its citizens in the host countries.

Hennebry and Preibisch (2012) recommend that for improving policy and practice in the management of temporary labour migration in agriculture, including greater autonomy for workers in choosing where they work and live, robust regulation of the recruitment process is required, together with the wider use of information sessions on health and safety, and access to certain settlement services such as basic language training. Though their research focused on temporary migration in Canada, they examined the temporary migration programme as a "best practice model" covering internationally recognized rights-based approaches through to labour migration, and providing some additional best practices for the management of temporary labour migration programmes.

They found that while the Canadian programme involves a number of successful practices, such as the cooperation between origin and destination countries, transparency in the admissions criteria for selection, and access to health care for temporary migrants, the programme does not adhere to the majority of best practices emerging in international forums, such as the recognition of migrants' qualifications, providing opportunities for skills transfer, avoiding imposing forced savings schemes, and providing paths to permanent residency.

The necessity of cooperation

Both sending and receiving states have roles to play in ensuring migrant workers are protected, and, in the case of death, that the families of deceased workers have access to effective compensation and assistance. In addition, receiving states have greater influence in international processes than sending states, which ought to create greater responsibility to ensure such international instruments are implemented and enforced.

Domestically, receiving states control the aspects that determine how migrant workers are treated, and the protection they have access

DOI: 10.1057/9781137451187.0009

to Recruitment agencies can be a significant source of misinformation and exploitation. In addition to adopting policies and legislation to ensure accountability and transparency among recruitment agencies, the foreign missions of receiving states can play a greater role in countering dishonest recruitment agencies. Receiving states also influence the degree to which migrant workers are integrated into existing protection systems, such as health care and judicial systems, within their jurisdiction. Receiving states can determine standards for remuneration and working conditions for migrant workers, and ensure that these are enforced by those who employ migrant workers. This is particularly important for domestic workers, who to date have been considered largely outside the sphere of protection due to the private nature of their work. Generally, receiving states – particularly those that host large numbers of Bangladeshi workers – can play a much greater role in protecting migrant workers. However, this will require significant pressure from both domestic populations and the international community, as many of these initiatives come at a cost. Cooperation between sending and receiving states will be necessary to deal with these costs, and to ensure that new policies and legislation are effectively implemented and enforced.

Governments should consider the opportunity to work with experts in the field, such as policymakers, economists, and civil society organizations active on the ground, to meet these challenges. Such experts are all useful and important partners to work with in order to develop specific knowledge and expertise in dealing with such a delicate issue. The government should increase the budget available for the application of progressive policies, specifically ones related to the training of the labour force and the dissipation of information and knowledge.

International cooperation is not the only way to promote workers' legal protections. There are several international conventions regarding the protection of workers and migrants that have failed to be enforced. One of the most important is the International Convention on The Protection of the Rights of All Migrants and Members of their Families,[1] adopted by the General Assembly of United Nations in 1990. This convention ensures the protection of regular and irregular migrants. Bangladesh should replace its current legislation with this UN Convention, and integrate the judicial system with other relevant ILO conventions (Siddiqui, 2005).

DOI: 10.1057/9781137451187.0009

Cooperation with civil society and NGOs is essential to identify fake agencies that take advantage of illegal migrants by exploiting them or extorting huge sums of money for travel costs, insurance, visa and other documentation. Government and civil society on the ground can work together in developing strategies to protect citizens and guarantee accountable and transparent agencies. This strategy would work together with the establishment of funding institutions for migrant workers. A funding institution that acts like a bank could be set up in order to provide loans to migrants like Bangladesh's Probashi Kalyan Bank (PKB). This would help migrants avoid exorbitant rates of interest on their loans, and their repayments would return to the government, rather than the private sector or individuals.

Three specific issues need to be analysed in greater detail. First, workers are exposed to occupational hazards, illness and contagious diseases. The government should put in place specific programmes for health service for both outgoing migrants and returnees, including their families. Secondly, having appropriate skills can help migrants protect themselves from exploitation. The government should facilitate developing those skills. The third issue is the vulnerability of female migrants. Special skills training could be offered to potential female migrants to help them find employment in more reputable domestic labour markets, because market demand for female caregivers is high. All bans should be lifted, but they should be given opportunities to migrate safely.

One of the most problematic areas in the entire migration process is that many migrants do not have access to banks. Therefore, the unbanked are forced to keep their money with their employers or friends. There is no accountability, no guarantee that the money would be given back as needed. The most complicated dilemma surfaces when the migrant workers die. It is important that all migrant workers maintain a bank account. The government should facilitate a system so that they can deposit the money for the least amount of charges. In the case of a migrant's death, the survivors can easily withdraw the money.

The most effective actions, regarding remittance transfer, which the Bangladeshi government can implement, are the liberalization of the payment market, financial product design and increasing financial literacy. Since in Bangladesh, the network of micro finance institutions is widespread it would be prudent to partner them so that remittances could be delivered through an institution that is already understood and trusted.

DOI: 10.1057/9781137451187.0009

Bangladesh should abandon the idea of introducing sophisticated financial instruments in a society where experience and capacity are limited. It should adopt conventional and basic financial tools such as affordable savings plans or micro insurance. The appropriate knowledge about the financial system and assistance is necessary to encourage development and growth. Generally speaking, the more educated people are the more aware they are of the opportunities for financial service, however the majority of the population prefer to save their money at home rather than investing or putting it in a bank account. The government should try to increase financial literacy so people even in rural areas can understand the advantages of some financial products and reinforce the existing system.

The most important recommendation for the sending governments would be to establish a separate cell in consulate offices stationed in countries with high numbers of abuse complaints. This will give migrant workers an option to report their situation, and seek aid. The government has to concentrate on conducting research on a continuous basis, documenting the situation that the migrant workers face, and then use the results of those studies to formulate their policy agenda. Further research is needed on gender differences in remittance behaviour; relations among remittance, investments, fraud, tax evasion and financial flow; the dynamics of uses of remittance (Ullah, 2011, 2011b); and the debate about the role of remittance in increasing or decreasing inequality. We conclude by quoting the Director General of the IOM "The paradox is that a time when one in seven people around the world are migrants, we are seeing an extraordinarily harsh response to migration in the developed world". As we write the conclusion of this volume (November 2014), we have learned that in Bahrain many Bangladeshi migrant workers fell victim to violent torture perpetrated by employers and 60 of them took shelter in the consulate of Bangladesh. We hope that soon employers will consider migrant workers as human beings and not expendable commodities, and that the ordeals endured by migrant workers will end one day.

Note

1 http://www2.ohchr.org/english/law/cmw.htm

DOI: 10.1057/9781137451187.0009

References

Abdulazeez, Y., Bab, I. and Pathmanathan, S. (2011). Migrant Workers' Lives and Experiences Amidst Malaysian Transformations. *The Social Sciences*, Vol. 6(5), 332–343.

Adhikary, P., Keen, S. and Teijlingen, E. V. (2011). Health Issues Among Nepalese Migrant Workers in the Middle East. *Health Science Journal*, Vol. 5, 171–172.

Afsar, R. (2009). Unraveling the Vicious Cycle of Recruitment: Labour Migration from Bangladesh to the Gulf States. Working Paper 63. Geneva: ILO.

Afsar, R. (2011). Contextualizing Gender and Migration in South Asia: Critical Insights. *Gender, Technology and Development*, Vol. 15(3), 389–410.

Ahmed, H. S. (2007). Neglected Heroes. *The Daily Star*, Dhaka, 20 June.

Ahn, Pong-Sul (2005). Prospect and Challenges of Out-Migration from South Asia and its Neighboring Countries. *Labour and Development*, Vol. 11(1), June.

Allard, T. (2011). Malaysia Plans Amnesty for Illegal Migrants. *Sydney Morning Herald*, 8 June 2014. Available at: http://www.smh.com.au/world/malaysia-plans-amnesty-for-illegal-migrants-20110607-1fr7h.html, accessed 8 June 2011.

Al-Najjar, S. (2004). Women Migrant Domestic Workers in Bahrain. In S. Esim and M. Smith (Eds), *Gender and Migration in Arab States. The Case of Domestic Workers* (pp. 24–40). Beirut: International Labour Organization, Regional Office for Arab States.

DOI: 10.1057/9781137451187.0010

Alonso, J. A. (2011). Migración Internacional y Desarrollo:unaRevisión a la Luz de la Crisis, CDP Background Paper No. 11, July 2011, United Nations Development Policy and Analysis Division.

Amnesty International (2011). Saudi Arabia Executes Eight Bangladeshi Nationals. Available at: http://www.amnestyusa.org/news/news-item/saudi-arabia-executes-eight-bangladeshi-nationals, accessed May 2012.

Arab News. (2008). Kuwait Punishes Companies Accused of Labor Abuse. *Arab News*, 15 August 2008.

Arnold, F. and Shah, N. M. (1986). *Asian Labor Migration: Pipeline to the Middle East.* Boulder: Westview.

Asia Sentinel (2011). Saudis Behead an Indonesian Maid, 21 June 2011. Hong Kong.

Bangladesh Bank (2007). Central Bank of Bangladesh Economic Data: Country-wise Remittance Inflows Available at: http://www.bangladesh-bank.org/, accessed 20 January 2012.

Bangladesh News (2008). Dhaka Calls for Probe into Kuwait Incidents. Bangladesh to the Gulf States. Working Paper 63, 6 August 2008. Geneva: ILO.

BBS (Bangladesh Bureau of Statistics) (2010). Report of the Household Income and Expenditure Survey 2010, Dhaka: Statistics Division, Ministry of Planning.

Bdnews24.com. (2014). URL bdnews24.com. Dhaka.

BMET. (2008). Bureau of Manpower, Employment and Training Statistical Report. Dhaka, Bangladesh.

BMET. (2009). Statistical Reports. Country-wise Overseas Employment (major Countries) in 2010. Available at: http://www.bmet.org.bd/BMET/stattisticalDataAction, 28 December 2011.

BMET. (2010). Bureau of Manpower, Employment and Training Statistical Report. Dhaka, Bangladesh.

BMET. (2011). Bureau of Manpower, Employment and Training Statistical Report. Dhaka, Bangladesh.

BMET. (2012). Bureau of Manpower, Employment and Training Statistical Report. Dhaka, Bangladesh.

BMET. (2014). Bureau of Manpower, Employment and Training Statistical Report. Dhaka, Bangladesh.

BOESL. (2011). Bangladesh Overseas Employment Services Limited Website. Available at: http://www.boesl.org.bd/, accessed.

DOI: 10.1057/9781137451187.0010

Buchenau, J. (2008). Migration, Remittances and Poverty Alleviation in Bangladesh Bureau of Manpower, Employment and Training (BMET), 2007. Country-Wise, Year-Wise Remittance. Report. Dhaka.

CARAM Asia, Human Rights Watch. (2011). Protecting Asian Migrants' Rights, 19 April 2011.

Castles, S. (2004). Why Migration Policies Fail? *Ethnic and Racial Studies*, Vol. 27, 205–227.

Castle, S. and Miller, M. J. (1998). *The Age of Migration: International Population Movements in the Modern World*. London: Macmillan.

Castles, S. and Miller, M. (2009). Migration in the Asia-Pacific Region. Migration Policy Institute (MPI) Report. Washington: MPI.

CESCR. (2004). Annual Report on Thirty-third Session of The Committee on Economic, Social and Cultural Rights: Chile.

Charles, H. (2009). Malaysia is No. 2: 82 Bangladeshi migrants died of "heart attack" this year. Charles Hector On Human Rights, Justice And Peace Issues, Labour Rights, Migrant Rights, For The Abolition of The Death Penalty, Towards an end of Torture, Police Abuses, Discrimination, 15 May 2009.

Chia, S. Y. (2006). Labor Mobility and East Asian Integration. *Asian Economic Policy Review*, Vol. 1(2), 349–367.

Chowdhury, K. R. (2008). Abuse of Migrant Workers Starts with Passport Seizure. Bdnews24.com, 6 August 2008.

Collinson, S. (2009). The Political Economy of Migration Process – An Agenda for Migration Research and Analysis. International Migration Institute Working Paper 12/2009. Oxford: IMI, James Martin 21st Century School.

Council of Europe. (2000). *Macrosocial Determinants of Population Health*, Sandro Galea (Ed.). Springer.

Daily Naya Diganta, (2009). Available at: www.dailynayadiganta.com, accessed 18 July 2009.

The Economist (2011). *The Future of Mobility*. Available at http://www.economist.com/node/18741382 Downloaded on 21 December 2014

Federation Internationale des Liges des Droits de l'Homme (FIDH) and the Egyptian Organisation for Human Rights (EOHR). (2003). Migrant Workers in Saudi Arabia. Available online at: http://www.fidh.org/IMG/pdf/sa0103a.pdf downloaded on 15 November 2014.

de Haas, H. (2009). *Mobility and Human Development*. New York: UNDP.

Frantz, F. (2008). Of Maids and Madams. *Critical Asian Studies*, Vol. 40(4), 609–638.

Gardner, K. (1995). *Global Migrants, Local Lives: Travel and Transformation in Rural Bangladesh*. Clarendon Press: Oxford Studies in Social and Cultural Anthropology.

Go, S. (2007). *Asian Labor Migration: The Role of Bilateral Labor and Similar Agreements' Regional Informal Workshop on Labor Migration in South East Asia*, http://www.fes.org

Godfrey, M., Ruhs, M., Shah, N. and Smith, M. (2004). Migrant Domestic Workers in Kuwait: Findings Based on a Field Survey and Additional Research. In S. Esim and M. Smith (Eds), *Gender and Migration in Arab States. The Case of Domestic Workers* (pp. 42–63). Beirut: International Labour Organization, Regional Office for Arab States.

Grant, S. (2005). International Migration and Human Rights. *Global Commission on International Migration*, pp. 7 30.

Halabi, R. (2008). Contract Enslavement of Female Migrant Domestic Workers in Saudi Arabia and the United Arab Emirates. *Human Rights and Human Welfare*, pp. 43–45.

Hansen, Randall (2010). An Assessment of Principal Regional Consultative Processes on Migration. *IOM Migration Research Series*, #38. Geneva: IOM.

Hansen, E. and Martin, D. (2003). Health Issues of Migrant and Seasonal Farm Workers. *Journal of Health Care For The Poor And Underserved*, Vol. 14(2), 153–164.

Haque, M. S. (2012). Migration Management Approaches and Initiatives in South Asia. In Md. Mizanur Rahman and AKM Ahsan Ullah (Eds), *Asian Migration Policy: South, South East and East Asia*. New York: Nova Publishers.

Hector, C. (2011). Migrants and Rights in Malaysia. Available at: http://charleshector.blogspot.com/2011/04/migrants-and-rights-in-malaysia-742006.html, accessed 23 January 2012.

Hennebry, J. and Preibisch, K. (2012). A Model for Managed Migration? Re-Examining Best Practices in Canada's Seasonal Agricultural Worker Program. *International Migration*, Vol. 50(1), e19–40.

Human Rights Watch. (2004). *Bad Dreams: Exploitation and Abuse of Migrant Workers in Saudi Arabia*. Available at: http://www.hrw.org/sites/default/files/reports/saudi0704.pdf, accessed 12 June 2010.

DOI: 10.1057/9781137451187.0010

Human Rights Watch. (2006). Building Towers, Cheating Workers: Exploitation of Migrant Construction Workers in the United Arab Emirates, HRW Report.

Human Rights Watch. (2007). *Exported and Exposed Abuses against Sri Lankan Domestic Workers in Saudi Arabia, Kuwait, Lebanon, and the United Arab Emirates.* New York: Human Rights Watch.

Human Rights Watch. (2008). *Challenging the Red Lines. Stories of Rights Activists in Saudi Arabia.* Available at: http://www.europarl.europa.eu/meetdocs/2009_2014/documents/darp/dv/darp20140122_05_/darp20140122_05_en.pdf, accessed 21 September 2010.

Human Rights Watch. (2009). *The Island of Happiness. Exploitation of Migrant Workers on Saadiyat Island, Abu Dhabi.* New York: Human Rights Watch.

Human Rights Watch. (2010a). *Rights on the Line. Human Rights Watch on Abuses Against Migrants in 2010.* New York: Human Rights Watch.

Human Rights Watch. (2010b). *Walls at Every Turn. Abuse of Migrant Workers Through Kuwait's Sponsorship System.* New York: Human Rights Watch.

ILO. (2011a). Global and Regional Estimates on Domestic Workers. Domestic Work Policy Brief 4. Conditions of Work and Employment Programme (TRAVAIL). Geneva: ILO.

ILO. (2011b). Conditions of Work and Employment Programme. Lebanon – Minimum Wages – 2011. ILO.

ILO. (2011c). Conditions of Work and Employment Programme. United Arab Emirates – Minimum Wages – 2011. ILO.

International Organization for Migration (IOM). (2002). Recruitment and Placement of Bangladeshi Migrant Workers: An Evaluation of the Process. Available at: http://www.iom.org.bd/publications/15.pdf, accessed.

IOM. (2003). Defining Migration Priorities in an Interdependent World, *Migration Policy Issues*, Vol. 1, 1–6.

IOM. (2004). Valuing Migration: The Costs, Benefits, Opportunities and Challenges of Migration. 88th section, MC/INF/276.

IOM. (2005). *World Migration 2005 Costs and Benefits of International Migration.* Geneva: IOM.

IOM. (2008). World Migration Report, Geneva: IOM.

IOM. (2009). World Migration Report, Geneva: IOM.

IOM. (2010). World Migration Report, Geneva: IOM.

IOM. (2011). Website. Available at: http://www.iom.org.bd/, accessed.

DOI: 10.1057/9781137451187.0010

Islam, S. (2012). *Death of Bangladeshi Migrants in the Countries of Destination*. Dhaka: The Daily Prothom Alo.

Jarallah, Y. (2009). Domestic Labor in the Gulf Countries. *Journal of Immigrant and Refugee Studies*, Vol. 7(1), 3–15.

Adams Jr., R. H. and Page, J. (2005). Do International Migration and Remittances Reduce Poverty in Developing Countries? *World Development*, Vol. 33(10), 1645–1669.

Jureidini, R. (2004). Women Migrant Domestic Workers in Lebanon. In S. Esim and M. Smith (Eds), *Gender and Migration in Arab States. The Case of Domestic Workers* (pp. 64–85). Beirut: International Labour Organization, Regional Office for Arab States.

Kader. A. (2008). Minimum wage means we can no longer afford to hire maids, *Gulf News*, 5 March.

Kapiszewski, A. (2006). Arab Versus Asian Migrant Workers in the GCC Countries. Paper presented at the United Nations Expert Group Meeting on International Migration and Development in the Arab Region, May 15–17, Beirut, Lebanon.

Kassim, A. (2005). Illegal Migrants and the State in Sabah: Conflicting Interests and the Contest of Will. In A. Kassim (Ed.), *State Responses to the Presence and Employment of Foreign Workers in Sabah*, Universiti Malaysia Sabah Press, Kota Kinabalu, pp. 1–36.

Khondker, H. and Rahman, M. (2009). *Bangladeshi Labor Migration to East and Southeast Asia* Report. Unpublished.

Kibria, N. (2004). Returning International Labor Migrants from Bangladesh: The Experience and Effects of Deportation. Working Paper # 28.

King, R. (2012). *Theories and Typologies of Migration: An Overview and a Primer*. Willy Brandt Series of Working Papers in International Migration and Ethnic Relations.

Knights, M. (1996). Bangladeshi Immigrants in Italy: From Geopolitics to Micro Politics. Transactions, Institute of British Geographers NS21, 105–123.

Kutasi, G. (2005). Labour Migration and Competitiveness in the European Union. *Transition Studies Review*, Vol. 12(3), 512–526.

Malecki, E. J. and Ewers, M. C. (2007). Labor Migration to World Cities: With a Research Agenda for the Arab Gulf. *Progress in Human Geography*, Vol. 31(4), 467–484.

Malik Nesrine (2011). Dubai's skyscrapers, stained by the blood of migrant workers. Available at: http://www.theguardian.com/commentisfree/2011/may/27/dubai-migrant-worker-deaths, accessed 27 May 2011.

DOI: 10.1057/9781137451187.0010

Mamun, K. A. and Hiranya, K. N. (2010). Workers' Migration and Remittances in Bangladesh. Availale at: http://www.shsu.edu/%7Etcq001/paper_files/wp10-02_paper.pdf, accessed 24 July 2011.

Marger, M. N. (2007). *Social Inequality: Patterns and Processes* (4th ed. rev.). New York: McGraw-Hill.

Marriot, R. (2008). Kuwait: Bangladeshi migrants export class struggle. *Libcom News*, 9 August.

Martin, S. F. (2001). New Issues in Refugee Research. Working Paper No. 41, Global Migration Trends and Asylum, Georgetown University, Washington.

Massey, D. S. (2007). International Migration in a Globalizing Economy. Available at: www.great decisions.org, accessed 18 May 2013. Malmö: Malmö Institute for Studies of Migration, Diversity and Welfare (MIM) Malmö University.

Massey, D. S. (2009). The Political Economy of Migration in an Era of Globalization. In S. Martinez (Ed.), *International Migration and Human Rights – The Global Repercussions of US Policy*. California: California University Press.

Massey, D. S., Joaquin, A., Graeme, H., Kouaouci, A., Pellegrino, A. and Taylor, J. E. (1993). Theories of International Migration: A Review and Appraisal. *Population and Development Review*, Vol. 19(3) (September), 431–466.

Maxwell Stamp Ltd. (2010). Study on the International Demand for Semi–skilled and Skilled Bangladeshi Workers, Dhaka.

MEWOE. (2011). Ministry of Expatriates Welfare and Overseas Employment Website. Available at: http://probashi.gov.bd/introduction, accessed.

Mittendorff, C. (2006). Victimization of Female Migrants from Bangladesh. Refugee Watch Online, 18 May.

MOHR. (2010). Annual Report of Ministry of Human Resources. Malaysia.

MOSAL (Ministry of Social Affairs and Labour). New Private Sector Labour Law, Act No. 6 of 10 February 2010, amending Law No. 38 of 1964.

National Rural Health Care Association. (1986). *The Occupational Health of Migrant and Seasonal Farm Workers: Report Summary*. Kansas City, MO: National Rural Health Care Association.

Nepali Mountain New (NMN). (2010). Nepali Women Involved in Prostitution in Kuwait: Report. Nepali Mountain New.

DOI: 10.1057/9781137451187.0010

OKUP (OvibashiKarmiUnnayan Program). (2009). Bangladesh Country Report. Part of Series 'HIV Vulnerabilities Faced by Women Migrants. From Bangladesh to the Arab States' of the UNDP Regional HIV and Development Programme. Shampur, Bangladesh: OKUP.

Olivier, M. and A. Govindjee (2013). Labour Rights and Social Protection of Migrant Workers: In Search of a Co-Ordinated Legal Response. Paper presented at the Inaugural Conference of the Labour Law Research Network (LLRN), Barcelona, Spain, 13–15 June.

Omar, M. A. (2005). Migration of Indonesians into Malaysia: Implications on Bilateral Relations. In A. Kassim (Ed.), *State Responses to the Presence and Employment of Foreign Workers in Sabah*. Universiti Malaysia Sabah Press, Kota Kinabalu, pp. 115–140.

Orozco, M. (2010). *Migration, Remittance and Assets in Bangladesh: Considerations about their Intersection and Development Policy Recommendations.* Inter-American Dialogue, available at http;//www.thedialogue.org/page.ofm?page ID=32andpubID=2372.

Palma, P. (2008). 5000 Bangladeshi Workers Go on Strike in Kuwait. *The Daily Star*. Available at: http://thedailystar.net/newDesign/news-details.php?nid=46854, accessed 22 July 2008.

Palma, P. (2011). Migrants Unprotected. *The Daily Star*. Available at: http://www.thedailystar.net/newDesign/news-details.php?nid=182327, accessed 20 April 2011.

Payne, R. (2005). *A Framework for Understanding Poverty* (4th ed.). Highland, TX: Process, Inc.

Peters, D. (2005). Cross Border Migration in the Context of Malaysia–Philippines Relations. In A. Kassim (Ed.), *State Responses to the Presence and Employment of Foreign Workers in Sabah*, Universiti Malaysia Sabah Press, Kota Kinabalu, pp. 141–161.

Piyasiri, W. (2002). *Asian Labour Migration: Issues and Challenges in an Era of Globalization*, International Migration Papers No. 57E, Geneva: ILO.

The Protection Project (2013). 100 Best Practices in the Protection of Migrant Workers. The Protection Project: The Johns Hopkins University Paul H. Nitze School of Advanced International Studies. A Series of 100 Best Practices, Volume II

Prameshwari, P. (2010). Ministry Probes Brutal Death of Indonesian Maid in Kuwait. *Jakarta Globe* 23 July 2010.

Prothom Alo. (2009). Six Bangladeshis Burned Alive in their Work Place in Saudi Arabia, 10 July. Dhaka: Prothom Alo.

DOI: 10.1057/9781137451187.0010

Prothom Alo. (2014). Soi Bosore chouddhohazar probashir lash asese (14 thousand dead bodies have arrived in six years), 10 February, Dhaka.

Rahman, M. (2009). Temporary Migration and Changing Family Dynamics: Implications for Social Development. *Population, Space and Place Special Issue: Rethinking the Migration–Development Nexus – Bringing Marginalized Visions and Actors to the Fore*, Vol. 15(2), 161–174.

Rahman, S. (2010). Human Rights Forum: Climate Change and Human Rights in Bangladesh. *Anthropology News*, Vol. 51(6), 30–31.

Rahman, A. (2010). *Migration and Human Rights in the Gulf. Viewpoint: Migration and the Gulf*. Washington: The Middle East Institute.

Rahman, M. (2011). Bangladeshi Migrant Workers in the UAE: Gender-Differentiated Patterns of Migration Experiences. *Middle Eastern Studies*, Vol. 47(2), 395–411.

Rahman, M. and Ullah, AKM. A. (Ed.) (2012). *Migration Policies in Asia: South, South East and East Asia*. New York: Nova Science.

Rajab, N. (2010). Migrant Workers and the Death Penalty in Bahrain and Saudi Arabia. Bahrain Center for Human Rights and CARAM Asia. Available at: http://www.bahrainrights.org/en/node/3054, accessed 5 November 2012

Rashid, M. (2011). How Effective are Bangladeshi Diplomats Abroad? *The Financial Express*, 6 April, Dhaka.

Riedel, E. (2007). The Human Right to Social Security: Some Challenges. In E. Riedel (Ed.), *Social Security as a Human Right – Drafting a General Comment on Article 9, ICESCR – Some Challenges*. The Netherlands: Springer.

Rosario, T. D. (2008). *Best Practices in Social Insurance for Migrant Workers: The Case of Sri Lanka*. Bangkok: ILO.

Sabban, R. (2004). Women Migrant Domestic Workers in the United Arab Emirates. In S. Esim and M. Smith (Eds), *Gender and Migration in Arab States. The Case of Domestic Workers*. Beirut: International Labour Organization, Regional Office for Arab States, pp. 86–107.

Shah, N. M. (2009). Arab Labour Migration: A Review of Trends and Issues. *International Migration*, Vol. 32(1), 3–28.

Shaham, D. (2008). Foreign Labor in the Arab Gulf: Challenges to Nationalization, *al Nakhlah*, 1–14.

Shariful, H. (2012). *Eight Thousand Dead Bodies in Three Years*. Dhaka: Prothom Alo, 25 May.

DOI: 10.1057/9781137451187.0010

Shaw, J. (2010). From Kuwait to Korea: The Diversification of Sri Lankan Labour Migration. *Journal of the Asia Pacific Economy*, Vol. 15(1), 59–70.

Siddiqui, T. (2003). Migration as a Livelihood Strategy of the Poor: The Bangladesh Case, Working Paper C1, Refugee and Migratory Movements Research Unit, Dhaka University, Bangladesh.

Siddiqui, T. (2005). International Labour Migration from Bangladesh: A Decent Work Perspective, Policy Integration Department National Policy Group, Working Paper No. 66, International Labour Office Geneva.

Siddiqui, T. (2009). Migrant Workers' Remittances to Bangladesh: Implications of GlobalRecession. Lecture delivered at Bangladesh Institute of International and Strategic Studies (BIISS), 23 April, Dhaka (The slides are available at http://www.biiss.org/tasnem.pdf).

Sikder, M. J. U. (2008). Bangladesh. *Asian and Pacific Migration Journal*, Vol. 17(3–4), 257–275.

Silvey, R. (2004). Transnational Domestication: State Power and Indonesian Migrant Women in Saudi Arabia. *Political Geography*, Vol. 23, 245–264.

Silvey, R. (2006). Consuming the Transnational Family: Indonesian Migrant Domestic Workers to Saudi Arabia. *Global Networks*, Vol. 6(1), 23–40.

Socialist World (2005). Oil Rich Arab States are Heaven for Capitalists, But Hell for Workers. 27 April, Available at http://www.socialistworld.net/doc/1710, Downloaded on 22 October 2014

Sri Lankan Bureau of Foreign Employment (SLBFE) (2009). SLBFE Socialistworld. 2005. Middle East: Immigrant workers forced to work without being paid for months. Available at: http://www.socialistworld.net/doc/1710, accessed 27 April 2011.

Susan, F. M. (2001). Global Migration Trends and Asylum, Institute for the Study of International Migration, Working Paper No. 41, Georgetown University, Washington DC, United States, *The Economist*, 2011. Available at: http://www.economist.com/node/21523188, accessed 8 November 2008.

The Daily Star. (2008). Immigration Department Turned to Business Centre: Anything Can be Done One Wishes by Bribes, 20 July, Dhaka.

The Financial Express. (2009). Bangladesh's Migration Costs Higher than Officially Allowed: WB, 16 August. Available at: http://www.thefinancialexpress-bd.com/2009/08/16/76285.html, accessed.

DOI: 10.1057/9781137451187.0010

United News of Bangladesh (UNB) (2010). *Bangladesh Faces Tough Times in Labour Markets Dhaka Mirror*, July 2010, Dhaka.

US Department of State. (2010a). Trafficking in Persons, Report 2010. Available at: www.state.gov/documents/organization/142979.pdf, accessed.

US Department of State. (2010b). Bahrain Human Rights Report 2010. Available at: http://www.ilo.org/dyn/travail/travmain. sectionReport1?p_lang=enandp_structure=1andp_sc_id=1andp_countries=REG2, accessed 28 October 2011.

US Department of State. (2010c). Jordan Human Rights Report 2010. Available at: http://www.ilo.org/dyn/travail/travmain. sectionReport1?p_lang=enandp_structure=1andp_sc_id=1andp_countries=REG2, accessed 28 October 2011.

US Department of State. (2010d). Lebanon Human Rights Report 2010. Available at: http://www.ilo.org/dyn/travail/travmain. sectionReport1?p_lang=enandp_structure=1andp_sc_id=1andp_countries=REG2, accessed 28 October 2011.

Ullah, AKM. A. (2003). Empowerment of Women in Bangladesh: Do NGO Interventions Matter? *Empowerment*, Vol. 10, 21–32.

Ullah, AKM. A. (2008). The Price of Migration from Bangladesh to Distant Lands: Narratives of Recent Tragedies. *Asian Profile*, Vol. 36(6), 639–646.

Ullah, AKM. A. (2009). Changing Governance in Population Migration: Theories and Practices in South and Southeast Asia. *BIISS Journal*, Vol. 30(1), 78–99.

Ullah, AKM. A. (2010a). *Rationalizing Migration Decisions: Labour Migrants in South and South-East Asia*. Aldershot: Ashgate.

Ullah, AKM. A. (2010b). Pre-marital Pregnancies Among Migrant Workers: A Case of Domestic Helpers in Hong Kong. *Asian Journal of Women's Studies*, Vol. 16(1), 62–90.

Ullah, AKM. A. (2011). Dynamics of Remittance Practices and Development: Bangladeshi Overseas Migrants. *Development in Practice*, Vol. 21(8), 1153–1167.

Ullah, AKM. A. (2011a). Forced or Development Induced Displacement? Occupied Palestinian Territories and International Conscience. *Journal of Internal Displacement*, Vol. 1(1), 5–17.

Ullah, AKM. A. (2011b). Dynamics of Remittance Uses and Development: Bangladeshi Labor Migrants in Hong Kong and Malaysia. *Development in Practice*, Vol. 21(8), 1153–1167.

DOI: 10.1057/9781137451187.0010

Ullah, AKM. A. (2012). *Divergence and Convergence in the Nation State: The Roles of Religion and Migration*. New York: Nova Science.

Ullah, AKM. A. (2012a). Bangladeshi Migrant Workers in Hong Kong: Adaptation Strategies in an Ethnically Distant Destination. *International Migration*, Vol. 51(2), 165–180.

Ullah, AKM. A. (2013). Mother's Land and Others' Land: "Stolen" Youth of Returned Female Migrants. *Gender, Technology and Development*, Vol. 17(2), 159–178.

Ullah, AKM. A. (2013a). Exposure to Remittances: Theoretical and Empirical Implications for Gender. *Journal of International Migration and Integration (JIMI)*, Vol. 14(3), 475–492.

Ullah, AKM. A. (2014). *Refugee Politics in the Middle East and North Africa: Human Rights, Safety, and Identity (Global Ethics)*. London: Palgrave Macmillan.

Ullah, A. K. M. Ahsan and Hossain, M. A. 2013, (2014). ' When Money Follows the Corpse: Remittances of Deceased Migrants in South Asia. In Md Mizanur Rahman, Tan Tai Yong and AKM Ahsan Ullah (Eds), *Migrant Remittances in South Asia*. London: Palgrave MacMillan, pp. 218–235.

Ullah, AKM. A. and Hossain, A. M. (2013a). Remittances of Deceased Migrants in Bangladesh. In T. T. Yong, M. Rahman and AKM. A. Ullah (Eds), *Migrant Remittances in South Asia: Social, Economic, and Political Implications*. London: Palgrave McMillan.

UNDP. (2002). Recruitment and Placement of Bangladeshi Migrant Workers: An Evaluation of the Process. Prepared for International Organization for Migration (IOM) Regional Office for South Asia, Dhaka. Prepared for Refugee and Migratory Movement Research Unit (RMMRU).

United Nations Population Fund (2006). UNFPA Annual Report. UNFPA: New York.

Wall Street Journal. (2011). Indonesia Stops Sending Maids to Saudi Arabia. US.

Wickramasekara, P. (2006). Labour Migration and Rights of Migrant Workers. Paper presented at the International Symposium on a Culture of Peace: Intercultural Understanding and Human Rights Education, 25–27 October, Seoul, Republic of Korea.

Wickramasekara, P. (2009). Labor Migration Policies and Practices in South Asia. Paper presented at the International Conference on

DOI: 10.1057/9781137451187.0010

Migration, Remittances and Development, Institute of Policy Studies of Sri Lanka.

World Bank. (1981). Labor Migration from Bangladesh to the Middle East. World Bank Staff Working Paper No. 454. Washington, DC: The World Bank.

World Bank. (2008). Migration and Remittances Factbook 2008. World Bank.

Yamanaka, K. and Piper, N. (2005). Feminized Migration in East and Southeast Asia: Policies, Actions and Empowerment. Occasional Paper No. 11. United Nations Research Institute for Social Development. Available at: http://www.unrisd.org/80256B3C005BCCF9/(httpAuxPages)/06C975DEC6217D4EC1257139 0029829A/$file/OP11%20web.pdf, accessed

DOI: 10.1057/9781137451187.0010

Index

DOI: 10.1057/9781137451187.0011

DOI: 10.1057/9781137451187.0011

DOI: 10.1057/9781137451187.0011

DOI: 10.1057/9781137451187.0011

DOI: 10.1057/9781137451187.0011

DOI: 10.1057/9781137451187.0011